PRAISE FOR

"*12 Fixes to Healthy* is one of the most comprehensive healthy lifestyle guides I've used. It's approachable and effective, no matter your age or health status. Each of the 12 Fixes provides easy-to-digest content and helpful tips to implement changes, but without being overwhelming. Instead, you're left with the belief that 1) you can change, 2) small changes matter, and 3) yes, healthy really can be this simple!"

CAROLYN WILLIAMS, PhD, RD, James Beard Award-Winner and Culinary Nutrition Expert, Author of *Meals That Heal*

"*12 Fixes to Healthy* utilizes current research and nutrition information in order to improve lifestyle habits from eating to sleeping to exercising. Encouraging the "fixes" to be done one at a time allows time for the reader to develop healthy lifestyle habits. No one has perfect habits, but *12 Fixes to Healthy* can help improve 12 important ones that many folks should be doing regularly. Judes provides an action plan for each of the 12 fixes to help readers develop healthier habits."

TOBY AMIDOR, MS, RD, CDN, FAND, Award-winning nutrition expert and Wall Street Journal best selling cookbook author

"If you want to eat better and move more but find these changes overwhelming, this book is for you. In her updated book, Judes walks you through—step-by-step—how to make these healthy changes part of your everyday life. Chock full of valuable tips and mouth-watering recipes, this book is ideal for anyone seeking to improve their health or for those who wish they could bring a dietitian home with them."

RIMA KLEINER, MS, RDN, LDN, Media Spokesperson, Company Consultant, Nutrition Coach at Smart Mouth Nutrition

"Judes identifies a unique non-diet approach to help the reader enjoy food without guilt. This book is a winner!

A nutrition expert writes it. Judes is a Registered Dietitian in the field for over 30 years. She has the track record and know-how to sourcing the most credible nutrition information chock full of research and scientific backing—no quack science here. If proof is in the pudding, just check out the bibliography!

This book is holistic, well written, scientifically evidence-based, and, most importantly, it's content is easy to digest and fun to navigate."

EMMA FOGT, MBA, MS, RD, LDN, FAND, Author of *Having Your All*

"*12 Fixes to Healthy: A Wellness Plan for Life* offers a refreshing and realistic approach to creating a healthy lifestyle. Judes' 12 Fixes have been an instrumental piece for our individual programs, as well as our corporate wellness lunch and learn presentations."

NICOLE GOODRICH, MS, RDN, CHWC, President of Anderson's Nutrition, a multi-state wellness company

"Judes has compiled 12 of the most important "fixes" to help people understand how to take off fat from the body and feel better without being on a diet. This 12-Fix approach is not a fad but an actual lifestyle book.

The book discusses body fat versus weight (which I'm always a fan of emphasizing a healthy body fat percentage versus weight). The book also has an essential section on snacking and meal timing, which helps with controlling hunger levels and making better food choices. Judes explains how to lose fat, keep muscle, and burn more calories from eating. Eat to lose, right? That's my motto, too.

Overall, this book is great for general health and wellness with practical tools on how to live a healthy lifestyle and incorporate real food and whole foods into your practice. There are simple tips on how to plan and prepare food, how to grocery shop, what to look for when food shopping, and how to cook the food."

SARAH KOSZYK, MA, RDN, Nutrition coach, sports dietitian, international author, and spokesperson

"This concise and easy to understand 12 Fix Wellness Plan in Judes' book, *12 Fixes to Healthy*, has been invaluable to me. As a BSN, Board Certified, Registered Nurse of over 45 years, I learned more about food and nutrition than I ever did in nursing school or my career that followed! The information I gleaned in this scientifically-based, easy-to-read book is so motivating.

I am convinced that if I had been educated on this information 50 years ago, my life would be so much different now. However, I have found that it is not too late to incorporate these simple habit changes, even at this time in my life. I'm making better choices, taking better care of myself and my family, and improving my personal health."

ANITA BLEYL-OLSZOWKA, RN, BSN, Administrative Nurse for DMBA's Hawaii Operations

12 Fixes to Healthy

12 Fixes
to Healthy

A WELLNESS PLAN FOR LIFE

JUDITH SCHARMAN DRAUGHON, MS, RDN, LDN
NUTRITION EDUCATIONAL SOLUTIONS, LLC

Copyright © 2020 by Judith Scharman Draughon, MS, RDN, LDN

All rights reserved. This book or parts thereof may not be reproduced in any form, stored in any retrieval system, or transmitted in any form by any means (including electronic, mechanical, photocopy, recording or otherwise) without prior written permission of the author.

For information about special discounts available for bulk purchases, sales promotions, fundraising and educational needs, contact publisher or author via email.

Published by Nutrition Educational Solutions, LLC

judes@foodswithjudes.com

LinkedIn: Nutrition Educational Solutions (Foods with Judes)

Instagram: @foodswithjudes, Facebook: @foodswithjudes, Twitter: @foodswithjudes

This book presents the ideas of the author and is not intended to substitute for a personal consultation with a healthcare professional. If you have any special medical conditions, consult with your healthcare professional before starting the program contained in the book regarding possible modificationmodifications for yourself. The ideas, procedures, and suggestions contained in this book are not intended as a substitute for consulting with your physician. All matters require individual medical supervision. The author shall not be liable or responsible for any loss or damage allegedly arising from any information or suggestion in the book.

The author is not responsible for your health or allergy needs that may require medical supervision. The publisher is not responsible for any adverse reaction to the recipes contained in this book.

ISBN	978-0-578-23134-1 Paperback
	978-0-578-23135-8 Hardcover

Cover photo © [Africa Studio] / Adobe Stock. Photos on pages iv, 64 and 162 © Katie Smith. All other photos © Judith Scharman Draughon unless otherwise indicated.

Book design by Katie Smith.

Index by InkSmith Editorial Services.

Thanks to my sweet young-adult children for enjoying my recipes through the years, appreciating quality food, and supporting me with this project. I want to give special thanks to Katie Smith for designing this beautiful book to highlight my *12 Fixes to Healthy* plan.

Thanks to my supportive husband for his willingness to adopt a new lifestyle and to welcome the many benefits for his life. I'm grateful to work with such phenomenally caring people and skilled professionals as Traci Fisher and Natalie Stephens. Their input and willingness to collaborate have been immeasurable in completing this program. Thanks to all the dear people in my life, both personally and professionally, who encourage me and excitedly sample my food and are eager for wellness tips.

TABLE OF CONTENTS

INTRODUCTION 1
Less Is More 2
Small Is Big 2
The Science of Great Health 3
12 Fixes to Healthy 4
Start with One Fix 6
Getting Started 9

FIX 1 SWAP SUGAR 11
Whole Food, Super Strength 11
Let It Go! 13
Sugar Swap 14
Artificial Sweeteners 17
Fructose Fallacy 19
Soda and Other Sugar-Filled Drinks 20
New Label Language 20
Whole and Ancient Grains 22
Gluten-Free Grains 23
FODMAP What? 24
Swap Sugar **26**

FIX 2 PLAN PROTEIN 29
The Power of Protein 29
How Much Protein? 33
Protein for Fitness 36

ix

Protein Shopping Tour 36
Plan Protein **42**

FIX 3 BOOST GUT 45

Gut Microbes 45
Improve Gut Microbe Balance 48
Prebiotics 55
Supermarket Yogurt Section 57
Yogurt Prep 59
Boost Gut **61**

FIX 4 PUSH PRODUCE 65

Produce Potential 66
Produce Shopping 69
Tricky Labels 71
Genetically Modified Foods (GMOs) 73
Produce Prep 75
Kitchen Formula Prep 80
Push Produce **86**

FIX 5 EAT EARLY 89

Eat According To Your Circadian Clock 89
Burn More Fat While You Sleep 91
Benefits Beyond Weight Loss 93
Breakfast is the New Dinner 94
Eat, don't starve! 94
Keto Diet 95
Eat Early **100**

FIX 6 MOVE MORE 103

Choose Your Own Adventure 103
Move More by Traci Fisher 104
The "Move" in Move More 105
The "More" in Move More 109
FITT 110
Type of Movement 110
Time and Frequency of Movement 111
Intensity of Movement 112
Fuel to Move by Judes 113
Move More **118**

FIX 7 WATER WELL 121

Hydrate to Burn Fat 121
Hydrate To Think 123
How Much Water? 124

From Soda to Water 124
Energy Drinks 127
Kombucha 128
Kefir 130
Alcohol 131
Hydrate for Fitness 134
Water Well **137**

FIX 8 SHIFT SURROUNDINGS 139

Creating New Routines! 139
Rearrange 140
Resize 144
Plan Your Surroundings 146
Search Your Surroundings 151
Budget Cut 155
Shift Surroundings **158**

FIX 9 SWITCH FAT 163

Fat Facts 163
Omega-3 Fatty Acids 164
Monounsaturated Fats 170
Fat Chance 172
Let's Go Shopping! 174
Unscramble the Code to Buying Eggs 176
Go Fish: Buying Fish 178
Kitchen Prep Rally 181
Fish Cooking Ideas 184
Switch Fat **186**

FIX 10 CHANGE GRAINS 189

Whole Grains for More Daily Detox 189
How Many Whole Grain Servings? 193
A Whole Grain For Everyone 195
Fuel for Fitness 196
Buy Great Grains 197
Overly Processed Food Posing as Whole Food 200
Oats 201
Preparing Great Grains 201
Beat the Rice and Pasta Rut 202
Grain Bowl Options 204
Change Grains **206**

FIX 11 MUNCH MINDFULLY 209

Chew and Swallow 209
Digestion Boost 211
Hunger Signals 212

Mindful eating 213
New Way of Thinking 217
Don't Let Cravings Control You! 220
Munch Mindfully **223**

FIX 12 SLEEP SOUNDLY 225

Move More To Sleep Better 225
Snooze to Lose... 226
Food to Snooze 230
Operation Shut Down.. 233
Sleep Soundly.. **235**

SUCCESS 239

You've Got This! 240
12 Fixes To Healthy 241
Ready, Set, Go! 242

REFERENCES I

CONTRIBUTORS XVII

INDEX XIX

ADDITIONAL RESOURCES XXIX

INTRODUCTION

My devotion to health has never come above my love for food. Too often, people associate healthy cuisine with boring food that is time-consuming and doesn't satisfy you. It has long been my mission to design delicious and doable techniques to help you feel amazing!

The true cost of foods

We're all in a hurry, so I will show you how simple and fast healthy meals can be. I'll arm you with easy meal ideas so you know just what to grab as you are hurrying through the grocery store, trying to figure out what you can get on the table ASAP.

You just may feel so good after eating like this that you won't be lured back into eating processed convenience snacks and meals very often. When there's an occasion that you do get stuck eating processed unhealthy foods because nothing else is available, you'll be disappointed and will notice a lower energy and concentration level.

I will show you how to eat nutrition-packed food without spending an arm and a leg—how to spend your money buying food that your brain and body thrive on, instead of empty foods you don't even realize make you foggy, tired, and craving more.

I will reveal the mystery of processed, packaged foods once and for all. You will learn just how to eat to curb those cravings for processed foods and how to avoid them—without fighting your willpower. I'll provide you replacement food ideas just as convenient and tasty but much more satisfying.

You can't afford not to eat food that is good for you. But it isn't as hard as you think.

LESS IS MORE

Studies show that doing something, however small, is better than doing nothing. I've seen this become reality time and time again in my years of practice. Making small changes over a long period of time can make a much bigger difference in your health and life than big changes that you can't sustain. My plan does NOT require perfection—a small change you do 70 or 80 percent of the time is something to be proud of, not ashamed of!

This plan isn't about being perfect. Break the all-or-nothing dieting mentality that leaves you feeling tortured and discouraged.

No matter who you are or what your history is, you can change your health for the better. I've seen it happen every day in my career. The small, simple steps are what make the difference in the long run. Making a few small changes that you follow most of the time, for the rest of your life, is so much better and easier than making many larger changes for a short time. Less truly is more!

SMALL IS BIG

Habits are incredibly powerful. Sometimes habits can feel like your enemy, but I will teach you to make them your ally. People don't realize what an enormous difference healthy habits—or unhealthy habits—can make in their lives, long term.

Habits might look small when you're forming them by making choices on a day-to-day basis. But before you know it, a few months go by. Then a year. Then five. As time keeps passing, the difference in your quality of life—due to all those "insignificant" choices—becomes very apparent. In my time as a dietitian, I have seen so much regret in those who didn't take care of themselves, and it is my hope for you that you won't have to feel the same.

From my own life, I have learned the power of small consistent choices in improving how I feel. Years ago, my friend Lisa asked me to start lifting weights with her. Lisa was consistent, so I was fairly consistent. Once a week, we went to the gym and lifted weights for thirty-five minutes. And I thought, *Thirty-five minutes, one time per week?* I really didn't expect, or even notice, a big difference.

Until... a year later, when I went on a ski trip with my family. All of the usual aches and pains normally expected after a ten-year hiatus from skiing never arrived! I didn't get sore the next day, either. *Or the day after that.* I could work hard, play hard, and feel great.

Those thirty-five minutes of weight training once a week about 75 percent of the time had made a difference after all.

Habits are a two-sided coin

Habits can work for you, or they can work against you. Food works the same way. It can be your medicine, or it can be your poison. Seemingly small food habits can feed the harmful bacteria in your gut and cause big inflammatory problems over time that result in excessive body fat and chronic disease, or they can keep you healthy for years at the cellular level. Small actions over time make big impacts on your quality of life!

THE SCIENCE OF GREAT HEALTH

Our understanding of what it means to be healthy is evolving. Science is showing us new possibilities for health that we never considered before. The research is better and there is a lot more of it. In fact, new technologies have opened up exciting research on how the food we eat affects our health and weight via the trillions of microbes living in our guts.

Finding missing puzzle pieces

Even back in 2011, the *New England Journal of Medicine* published a detailed analysis[1] of several factors affecting the body weight of 120,877 people, helping piece together the puzzle of body weight.

Researchers followed participants for twelve to twenty years. What they discovered was that seemingly insignificant habits in eating and exercise resulted in large changes in body weight over the years. The average participant gained almost one pound every year over twenty years, while some lost weight during the same time.

The types of foods they ate made the biggest difference. The researchers found that the number of calories eaten didn't seem to matter as much as the types of calories that were consumed. Fruits, vegetables, and whole grains were common among participants who lost weight and maintained healthy weight levels. Interestingly, those who ate yogurt lost the most weight: an average of 0.82 pounds every four years, which the researchers attributed to the yogurt's healthy bacteria, a newer concept back then. Those who ate nuts and nut butter were close behind, highlighting the fact that high-calorie foods can actually promote weight loss.

Exercise and sleep also affected their weight. Participants who both ate healthily and exercised gained 1.76 fewer pounds on average than those who didn't exercise.

Those who slept less than six hours or more than eight hours a night were typically the people who gained the most.

12 critical pieces

Some of these science-based strategies reported in this study were just emerging in the research back in 2011, but now the scientific research that supports them is extensive. These strategies, as well as other patterns prominent in the scientific literature, are built into my 12 Fixes wellness plan in this book to promote a healthier lifestyle.

Since this study and the creation of these 12 Fixes in 2015, scientific research continues to reinforce these pieces to the wellness and weight puzzle. In fact, it's amazing to me how these 12 Fixes support each other so well. Each fix helps you to follow the other fixes, indicating that these truly are the puzzle pieces we've been missing through the years. We're clearly getting closer and closer to wellness truth, and it's so exciting!

12 FIXES TO HEALTHY

Over the years in both practice and research, 12 elements have stood out as the most impactful when it comes to losing fat weight, reducing inflammation, and preventing disease. These 12 elements make up my *12 Fixes to Healthy: A Wellness Plan for Life*.

Food preferences

Don't worry if you have dietary restrictions, food sensitivities, or if you're vegan or you love meat, for instance. These 12 Fixes are flexible and fit into just about any lifestyle! Individual food limitations or preferences can easily be weaved into the 12 Fixes found in this plan.

Mediterranean diet base

Some of these fixes are about what to eat. The Mediterranean-style eating pattern is reflected in these fixes because the scientific research clearly shows the health benefits of a mostly plant-based diet rich in fruits, vegetables, beans, lentils, whole grains, herbs, nuts, seeds, olive oil, and avocados, along with seafood, cultured dairy (like yogurt and artisan cheese), and eggs.

How to eat

Two of these fixes are about how to eat, so no matter what you're eating you can make some simple changes that make a big difference to your health and weight. By the way, these two fixes allow you to get greater satisfaction from less food and without fighting your willpower. Yes, that's right, you eat less food without even realizing it and you are more satisfied.

When to eat

Even overnight intermittent fasting paired with breakfast and aligned with the circadian cycle can be integrated into this plan. Eating food at night conflicts with your nighttime pathways that are supposed to encourage sleep, while during the day your body metabolizes food most effectively. This plan offers the cells in your body a needed window of time to repair and restore during the night while your digestive system is quiet.

Non-food fixes

Fixes offering direction as to what to drink, how to move, and even sleep are also part of this wellness plan. These 12 Fixes start where you're at now and help you move to a healthier place, without the pain of a "diet" and with the enjoyment of guilt-free thinking.

Powerful fixes

These 12 fixes are the keys to burning fat, improving gut health, avoiding chronic disease, protecting your brain from cognitive decline, and becoming more productive and physically stronger. All these 12 Fixes enhance both your gut health and your brain health indirectly or directly. And the best part is you don't have to follow them to the letter 100 percent of the time.

Perfection not required

You'll still make a big difference in your health as long as you do these 12 Fixes "most of the time" from now on. Even doing these fixes some of the time will help your health and weight over the course of time.

Since perfection is not required (or needed), you never have to feel guilty. This may be a shift in the way you think about your food and your health. It's about feeling empowered by every healthy choice you make each day, even when it seems small. Scrap the mentality that you have to do it all. Track the things you do right, not the things you do wrong. Before you know it, guilt will disappear. Health and success will take its place.

Every fix you make, large or small, is going to have a positive impact on your health in the long run!

START WITH ONE FIX

The 12 Fixes are the most important actions to take to get the biggest bang for your effort. These little behaviors are simple, *sometimes seemingly insignificant*, and often these little changes are easy when taken on one at a time. However, their effects can be enormous, especially when even just a few of them are added together over a lifetime.

The 12 Fixes found throughout these pages may indeed look simple, but their effects are incredibly far-reaching. After just a couple of months of utilizing these fixes, you'll begin to feel more energized, awake, and astute. Your body mass will begin to shift from fat to lean tissue. Inches will drop from your waistline without you feeling hungry or feeling deprived. You'll feel more satisfied than you ever have before. And as you enjoy making healthier choices, you'll feel like you can conquer the world.

One fix at a time

This *12 Fixes to Healthy* plan is not like traditional weight-loss programs. It involves eating, not restricting. It boosts your metabolism, rather than slowing it down. It helps fat cells burn calories instead of breaking down lean muscle. This all happens by concentrating on one fix at a time!

This plan provides practical action steps and research-based strategies to help you follow each fix. You will celebrate successes instead of dwelling on failures. This plan turns seemingly huge challenges into easy-to-reach goals. You can transform your health for the better with this plan and makes it easy for you to keep it that way.

When working toward a new lifestyle starts to feel overwhelming, remember: just start with one fix.

FIX 1 SWAP SUGAR

Replace sugar, refined white flour, and processed foods with whole foods. Swap out unhealthy sweeteners such as sugar and high-fructose corn syrup with healthy sweeteners such as whole fruit, crushed fruit, raw honey, or pure maple syrup. Substitute refined flours with whole and ancient grains. Forgo commercially processed food in favor of foods in their whole, natural form as much as possible.

FIX 2 PLAN PROTEIN

Eat a total of 75 to 150 grams of protein spread throughout the day, with at least 30 grams of protein for breakfast. Protein spread throughout the day keeps you full, keeps your brain's fuel source steady, and helps increase muscle mass. The high-protein boost in the morning even helps to reduce your cravings for unhealthy foods in the late afternoon and evening, when willpower is weakest. Multiply your body weight in pounds by 0.6 to determine approximately how much protein you need to eat throughout the day.

FIX 3 BOOST GUT

Eat a fermented food, like yogurt, every day. Try to eat or drink food containing live active bacteria, such as lower-sugar yogurt, every day. The live active bacteria found in some foods boost your gut health. They don't stick around for long, so you need to invite them in on a daily basis. Consistency is more important than quantity; the more regularly you eat fermented (also referred to as cultured) foods, the more that the good bacteria in these foods can boost your gut health.

FIX 4 PUSH PRODUCE

Make fruits and vegetables 50 percent of your meals and snacks. Eating fruits and vegetables is your daily "detox cleanse," giving your immune system a tremendous boost. Produce in its whole-food form delivers thousands of natural plant chemicals that help cells repair DNA damage caused by pollution, poor diet patterns, alcohol, smoking, and other harmful toxins. Beans and lentils are vegetables too, so don't leave them out. However, french fries don't count.

FIX 5 EAT EARLY

Front-load your eating and avoid eating in the evening. Shift from eating during the evening to eating the majority of your food in the daytime. Eat to fuel your body during the day when you're concentrating and moving and your body handles carbohydrates the best. Avoid eating in the evening after dinner when your nighttime pathways that encourage sleep are in conflict with eating. Your body readily uses carbohydrates for fuel, and the sooner in the evening you use up the carbohydrates for energy, the sooner you go into fat-burning mode while you're sleeping. Eat breakfast like a healthy king, lunch like a prince, and dinner like a pauper.

FIX 6 MOVE MORE

Move your body more. Whatever movement you currently are doing, add a little more. Our amazing bodies are constantly learning and adapting. Moving more can mean adding more time, more frequently, more intensely, or even more variety and more fun! The more you move, the better. Each movement adds up to create optimal health.

FIX 7 WATER WELL

Drink water throughout the day and evening. Drinking water frequently and consistently ensures that your fat cells are swollen with water to burn fat more efficiently. Dehydrated cells are more likely to store fat instead of using it for energy. Of course, once the cells are completely hydrated, more fat won't be burned by drinking more water. But drinking water helps fat cells burn to their full capacity. Keep a glass of water within arm's reach at all times. If it's there, you're more likely to drink it!

FIX 8 SHIFT SURROUNDINGS

Rearrange, resize, plan, search your environment. Making some simple changes to your surroundings can trick your brain into wanting to eat less unhealthy food and to eat more health-promoting foods, no willpower needed. Pay attention to what signals around you are encouraging you to eat the wrong types of foods in the wrong amounts. I'll show you ways you can make some shifts to avoid the problems altogether. There are so many ways to make this happen. The list of small, seemingly insignificant yet effective strategies is extensive.

FIX 9 SWITCH FAT

Replace less-healthy fats with healthier fats. Replace lard, butter, shortening, and margarine with fats rich in monounsaturated fats such as olive oils, refined safflower oil, canola oil, and nut oils. Eat peanuts, nuts, seeds, and avocados on salads, replace mayo with avocado on sandwiches, and eat sustainable fish at least twice a week to consume more omega-3 fatty acids. Omega-3 fatty acids are also found in walnuts, flax seeds, and chia seeds.

FIX 10 CHANGE GRAINS

Eat between three and six servings of whole grains per day instead of refined carbohydrates. Make the switch and replace refined grains like white, refined bread and buns, white rice, and white-flour pasta with whole grains like whole-grain bread, black or brown rice, and whole-grain pasta. A serving of grain, such as brown and black rice, farro, barley, oats, quinoa, sorghum, or whole-grain pasta, is typically half a cup or one slice of whole-grain bread. More servings may be needed on days filled with intense exercise or to maintain your weight if you are not trying to lose weight. Most importantly, skip the processed grains.

FIX 11 MUNCH MINDFULLY

Chew small bites of food slowly and completely, only eat when you're hungry, and stop eating just before you get full. Take small bites of food and chew slowly until nothing solid is left before you swallow. The more you chew, the more nutrients you absorb, the calories you burn, and the better you digest your food. Instead of eating just because it's "time" to eat, eat only when you're truly hungry, and then stop eating just as you become satisfied rather than starting to feel uncomfortable. These simple tricks actually help you to eat less with more enjoyment!

FIX 12 SLEEP SOUNDLY

Sleep seven to nine hours every night. Set a time to go to bed, and stick to it. If you're not tired, read a book instead of watching TV to relax. If you get enough sleep, you'll be more productive the next day. Inadequate sleep encourages higher levels of hunger hormones and decreases our fat-burning and muscle-building activity during sleep. In other words, the better you manage your sleep, the better you'll be able to manage your weight. Snooze to lose!

GETTING STARTED

Start by taking your waist measurement, and remeasure exactly the same way every month. Access a body composition test to find out your body fat and lean tissue body percentage if available. Then retest every three months. However, if you don't have access to this test, no worries! Body measurements will suffice and are still a better indicator of your progress than the number on the scale.

Take advantage of the *12 Fixes to Healthy* app to help you adopt all 12 Fixes to your life.

Choose one fix

Choose only one fix to focus on for at least one week, but, ideally, for a month. After spending some time on your "focus fix," continue to follow and track it as you add new fixes each week or each month. After you have experienced and concentrated on all of the fixes, continue to follow all 12 Fixes, but concentrate on the harder fixes (one at a time) until all 12 become an integral part of your life.

Track

Track these fixes throughout the day. Each fix is different, so tracking is different. But it's also simple. If you are eating a cultured food each day for Fix 3, then you track that you ate it. Or you will track how many grams of protein you took in or if you ate half your meals and snacks as produce.

I intentionally choose not to count calories. Calories are like gas in a car; the quality of the gas affects how your body runs, especially in the long term. Your body uses different types of calories differently. At the end of the day, the total number of calories we eat is less important than the type of calories we take in. Despite plenty of research to support the idea that different types of calories are processed differently by the body, many dieters still view total calories as the best determining factor in losing weight.

The grocery shelves are full of highly refined carbohydrate one-hundred-calorie snack packs without any reference to the type of calories. Most apps and trackers count calories, perpetuating this distracting calorie-counting confusion. But not this app for this program. Each fix is tracked instead with the *12 Fixes to Healthy* app.

This isn't a diet. Yes, you're going to lose fat weight, but this is actually a lifestyle AND a new way of thinking.

Focus on your good choice, not your bad ones! Build new routines around the fixes, even if you don't follow them perfectly. Feel good about the positive changes you make, no matter how big or small they are!

Above all, remember: don't push yourself so hard that you get discouraged and give up. Cut yourself some slack, and give yourself as much positive reinforcement as possible. Perfection is not nearly as important as longevity!

FIX 1
SWAP SUGAR

It's so much easier to replace certain foods with delicious alternatives instead of trying to eliminate them. So instead of trying to get rid of sweets altogether, replace processed sugar and white flour with foods that are closer to their whole-food form. The best three whole food sweeteners are sweet gifts from Mother Nature: fruit, raw honey, and pure maple syrup.

It's still a mystery how the different nutrients in whole foods work together. More than twenty-five thousand phytochemicals have been identified in plant-based whole foods. One apple contains more than ten thousand natural plant chemicals[1] laboring together to boost your health and keep your body going.

So don't be surprised if you feel a lot better and have more energy with just this one fix to replace sugar, refined white flour, and processed foods with whole foods!

WHOLE FOOD, SUPER STRENGTH

Foods in their whole form, without processing or with minimal processing, are even better for your body than you realize.

Whole foods from Mother Nature do more than just make you feel great mentally and physically. They also help you to lose weight and to prevent a host of health problems.[2] For example, fruits and vegetables contain super-concentrated phytochemicals that deliver a major boost to your immune system. When you eat plenty of them instead of overly processed foods, not only do you lose weight and reduce your inflammation[3] but you also fight off illnesses, help lower your blood pressure, and reduce your risk of diseases[4] such as cancer, diabetes, and heart disease.[5]

Eat less, want less, weigh less

Eating foods close to their whole-food form can help you to be more satisfied and crave less food overall. Many large association studies have been reporting this conclusion for several years, but despite the grand numbers of people in these studies, it's difficult to know for certain that the conclusions are reliable.

However, a tightly controlled study recently published by the National Institutes of Health[6] showed an actual cause-and-effect link between processed food and the following: weight gain, less satisfaction, eating more quickly, and greater overall food consumption.

The study confirmed previous studies' findings that eating and drinking processed foods actually drives people to overeat and gain weight, compared with those eating mostly whole or minimally processed foods. Study participants on the ultra-processed diet ate an average of 508 more calories per day and ended up gaining an average of two pounds in a two-week period. People on the whole food and minimally processed diet, meanwhile, ended up losing about two pounds on average over a two-week period.

Significantly, the large calorie consumption and weight-gain differences between the two diets occurred even though the diets were equal in carbohydrates, fats, proteins, sugars, fiber, sodium, and carbohydrates. The minimally processed diet included a lot of fresh fruits, and the ultra-processed diet wasn't as junky as you

> **I SPY HIGHLY PROCESSED FOOD**
>
> To find out if a product is highly or overly processed, go straight to the ingredient list and use my two-step method.
>
> Can you buy these ingredients to use in your own kitchen? Or are they only made in a lab or using a chemical process (like hydrogenated oil, high-fructose corn syrup, soy protein isolate, or aspartame)? If you can't buy the ingredients to make this product at home, consider it highly or overly processed!
>
> Take the process a step further: Does the product use whole fruit, raw honey, or whole-grain flour, or does it use processed sugar, processed honey, or refined flour?

might think, containing typical convenience foods such as canned soups and canned chicken.

Weight gain, packaged food, and inflammation are all linked. Many of the foods Americans love to eat—highly processed with refined flours and sugars—are wreaking havoc in our bodies. Your body craves more processed carbs. This leads to more overall body fat, causing ongoing inflammation inside your body.

No weight issues

Even if you manage to keep your weight down while eating these foods, your body is still left with chronic inflammation and poor gut health. Both will decrease your energy and ability to concentrate and will eventually lead to illness of some kind down the road. Populations that adopt an American diet high in processed foods have more disease outcomes within a few years.

Your immune system can only fight off so much at once. Not only do processed foods deprive your body of valuable nutrients that help fight toxins, they add more problems for your body. All this takes a significant toll on your immune system.

Whole food and depression

There is some evidence that a poor diet, high in ultra-processed foods filled with refined carbs and less healthy fats, is associated with depression, while a healthy diet rich in vegetables, fruit, fish, and lean meat is associated with reduced risk of depression.[7] A controlled study in adults reported that diet intervention can reduce clinical levels of depression.[8]

A 2019 study involving young-adult college students reported reduced depression symptoms within three weeks. Almost eighty students who experienced elevated levels of depression symptoms and habitually ate an overly processed diet were randomly assigned to a three-week whole-food eating pattern (including fruits, vegetables, whole grains, and lean proteins) or to a control group eating a poor diet. The group eating a whole-food eating pattern had statistically significant decreased depression symptoms and improved mood compared to the control group in three weeks.[9]

LET IT GO!

America has a sweet tooth. We all know that we eat too much sugar and high-fructose corn syrup, but we keep eating it anyway because we've become accustomed—even addicted[10]—to the taste of processed sweets. Some studies have even reported that rats had a stronger response to sugar than to cocaine.[11]

Sugar has a rap sheet a mile long. It's empty of nutrients, and it's a concentrated source of calories that rush into our bloodstream too quickly, inviting a flux of insulin to handle the sugar rush. That insulin surge promotes fat cells to be set in fat-making gear instead of burning mode, trapping the calories to make fat. The body responds with hunger and cravings, especially for processed carbs, because those trapped calories aren't available for energy. This cycle repeats, leading to obesity[12] and inflammation as well as the many health problems that go along with them, including type 2 diabetes and heart disease.

The research overwhelmingly indicates that when we cut back on sugar and refined flour, many of those ill effects disappear.[13]

Brain on sugar

Even your brain benefits from this lower-sugar shift.[14] While it's true that your brain needs sugar to think, stay focused, and keep your attention from drifting, it's critical that the sugar is broken down by the body from whole foods rather than from concentrated sources of processed sugar. Without the thousands of nutrients in whole food, sugar is absorbed too quickly, leading to a burst of energy followed by a slump. The sugar contained naturally within whole foods, however, results in a more consistent, longer-lasting energy source for the brain.[15] You need to make a shift from processed, refined sugars to sweeteners found in their whole, natural-food forms. It's time to let it go and make the swap to healthier sweeteners!

SUGAR SWAP

The healthiest sweeteners for you are whole-food sweeteners that deliver vitamins, minerals, phytonutrients, and fiber along with their sweet taste. You should eat whole foods in their natural form, and sweeteners are no exception. Fruit in its whole form, raw unfiltered honey, real maple syrup, and dark chocolate or cocoa are all-natural sweeteners filled with fiber, other natural nutrients, and plant chemicals that help negate the negatives of the natural sugar in these foods.

Applesauce, mashed bananas, pears, mangos, or any pureed fruit, rather than processed sugar, can help to sweeten food. Whether it's in a smoothie or a dessert, fruit can help you cut down on processed sugar and processed honey.

You can replace half of the sugar in baked good recipes with fruit puree. Since they also contain water, reduce the liquid in a recipe by one-quarter of a cup for every cup of puree you add.

CINNAMON

Cinnamon mimics sweetness to help you cut back on sugar, keeping your blood glucose levels more steady. When cinnamon is added, less sugar is needed for your cup of coffee or your freshly made muffins. Consequently, less insulin is needed to handle the rush of sugar to your blood, which may keep your body out of fat-making mode.

Further, one teaspoon of cinnamon carries an antioxidant punch similar to a half a cup of blueberries. However, the claims that cinnamon reduces blood glucose levels in diabetics and decreases the risk of heart disease haven't yet been substantiated in the body of research.[16]

Whole-fruit sauce or syrup

Commercial pancake syrup is a processed food that many people have a hard time letting go of. Luckily, whole fruit makes a perfect substitute.

Top some whole-grain pancakes with warm, unsweetened applesauce doused with cinnamon for another way to use fruit instead of sugar for sweetness. Unsweetened applesauce or other mashed or pureed fruit also adds moisture to baked goods.

My famous blueberry sauce!

I've been making a quick fruit sauce with frozen berries and chia seeds for years. My kids grew up on it and still come home as young adults wanting this sauce instead of syrup on their pancakes, waffles, crepes, french toast, and plain yogurt. The fruit can just cook down to thicken, but the chia seeds speed up the whole process by absorbing the liquid in the sauce. These useful little seeds add lots of plant chemicals, like antioxidants, that naturally preserve the sauce and boost its nutrition. The seeds are soft and the fruit is dark, so no one even knows that chia seeds are in the sauce.

To make this delicious sauce, just throw a little bit of juice or water in a saucepan along with frozen blueberries, a tablespoon or two of chia seeds, and cinnamon. Cook it down while making pancakes or waffles. Add a little raw honey or real maple syrup if you want it a little sweeter. In ten minutes you have yummy, warm blueberry sauce!

Dessert dates

Dates are packed with vitamins and minerals, including B6, A, K, niacin, riboflavin, folate, potassium, magnesium, manganese, iron, zinc, and calcium. Each date you eat contains almost 2 grams of fiber, which slows down the absorption of sugar into the bloodstream. Dates are also rich in protein; they contain five times more protein than any other fruit. They are an excellent source of antioxidants, and they contain twenty-three amino acids. Dates are the most nutritious sweetener we have, and they are easiest to include when you are already using a high-powered blender for your recipe. Three-quarters of a cup of dates equals about the sweetness of one cup of sugar in a recipe. The more you can replace sugar with whole foods such as naturally sweet dates, the better it will be for your health.

Raw honey

Honey is also loaded with vitamins, minerals, phytonutrients, and living enzymes with antiviral and antibacterial properties[17]—but it needs to be raw! These valuable nutrients are destroyed during the heating and pasteurization process in commercially processed honey, which causes it to rush into the bloodstream just as sugar does.

The glycemic index (GI) is a ranking of carbohydrates in foods according to how they affect blood sugar levels. Carbohydrates with a GI value of 55 or less out of 100 are more slowly digested, absorbed, and metabolized, thus causing a lower and slower rise in blood sugar and, therefore, insulin levels. Lower insulin levels keep fat cells in burn mode rather than the store-and-grow mode.

In fact, processed sugar and processed honey have similar high glycemic index ratings, around 70 to 90. By contrast, raw honey has a glycemic index rating of only about 30 to 40, according to the Glycemic Index Database.[18] Not only does raw honey promote a more stable blood sugar level[19] but, because it's sweeter than other sugars, you can use about 25 percent less of it when sweetening your foods and drinks.

NO HONEY FOR BABY!

Never give any kind of honey to a child under one year of age. An infant's immune system can't deactivate the botulinum spores found in honey. These spores may grow in the digestive tract and produce toxins, which will make the infant sick.

Pure maple syrup

Pure maple syrup tapped from trees—not the commercial pancake syrups made from high-fructose corn syrup—is a healthy, natural sweetener. While the body breaks down the sugar in maple syrup a little faster than it does raw honey, it's still slower than table sugar or processed honey, with a glycemic index rating of about 50. It may not provide as many phytonutrients or as much fiber as crushed fruit does, but it's easy to use and adds a delicious flavor along with its sweetness.

> **COMMERCIAL PANCAKE SYRUP**
>
> Even "natural" pancake syrup or table syrup made without high-fructose corn syrup is not the same as pure maple syrup that comes from the sap of a tree. Pure maple syrup has one ingredient: maple syrup. Commercial pancake syrup has from ten to fifteen ingredients.
>
> This is not a green light to use pure maple syrup as you would fruit in its whole fruit form. You still need to use as little as possible, but enjoy your gift from nature to replace regular sugar. Always warm pure maple syrup to thin it out so that you don't need as much. It tastes better warm too!

Dark chocolate and cocoa

Cocoa can be eaten frequently. I even add cocoa to my chili. Homemade desserts made with cocoa are better for you if you use a more nutritious sweetener and fat source. Dark chocolate contains not only cocoa but also cocoa butter and sugar, so it needs to be limited to a smaller amount. However, a small amount of dark chocolate eaten regularly is a positive habit. Dark chocolate and cocoa are full of antioxidants (even more than kale and spinach) and may also have anticancer properties. They contain flavonoids, which boost heart health by increasing blood flow, lowering blood pressure, reducing bad cholesterol levels, and keeping arteries flexible.

The less processed your dark chocolate is, the better. Avoid chocolate where sugar is listed as the first ingredient. Aim for a 70 percent cocoa content or higher in your dark chocolate to get the greatest amount of disease-fighting flavonoids out of it. And remember not to overdo it. Stick to just a square or two a day.

ARTIFICIAL SWEETENERS

Some sweeteners start out healthy in nature but by the time you buy them, they have been highly processed. Agave nectar, for example, is presented as a healthy,

SODA AND OTHER CALORIC DRINKS

One 12 ounce can of soda contains close to 50 grams of added sugar. That is 13 teaspoons, surpassing the daily limit of nine teaspoons of sugar for males and six teaspoons for females. Soda contributes so much of the sugar adding to obesity and diabetes issues.

Drinks with calories (even healthy drinks) bring blood sugar levels up too quickly, and the sugar lingers for too long after the meal, requiring the need for more insulin to transport the sugar. This extra insulin prevents fat cells from burning as much energy. As a bonus to your waist, the research indicates that you don't instinctively replace the calories with extra food instead. So, drink water instead of soda. Sparkling water—with its fizz and no added sugar—can help you make this health-promoting transition.

natural sweetener because it doesn't stimulate insulin the way glucose does. However, agave nectar is 90 percent fructose! That's even more fructose than high-fructose corn syrup, which has a fructose content of 55 percent. Fructose in processed foods is absorbed quickly and ends up promoting dangerous belly fat and inflammation.

Although many artificial sweeteners have essentially no calories, they do affect the body. Our taste receptors perceive them to be hundreds to thousands of times sweeter than sugar, and those who consume these synthetic chemicals may find naturally sweet food less appealing. Artificial sweeteners may also increase insulin production, causing calories to be shifted to fat cells rather than burned for energy.[20]

GUT BALANCE AND ARTIFICIAL SWEETENERS

There also seems to be some evidence that diet sodas alter the gut microbiome balance to promote weight gain and type 2 diabetes despite their lack of calories.[21] More research is needed to solidify this idea, but it appears to have grounds for interesting investigation.

If you need to use an artificial sweetener, the one I recommend you use is monk fruit sweetener, and even then, it's almost always better to eat the fruit in its whole form. The monk fruit is an ancient Chinese fruit two hundred to three hundred times sweeter than sugar. However, by the time it reaches the grocery store shelf, it has still been refined and processed.

FRUCTOSE FALLACY

Fructose doesn't spike our blood sugars like glucose can. Consequently, processed sweeteners higher in fructose (like agave nectar) with a lower glycemic index appear to be a healthier choice. But processed fructose can promote body fat as well. In fact, fructose can cause dangerous fat that surrounds our organs, creating a fatty liver and disease-promoting belly fat.

High-fructose corn syrup

Realizing that it could replace fats with inexpensive refined carbohydrates back in the 1970s, the food industry has since dramatically increased the use of high-fructose corn syrup in foods. Processed, packaged food became the way of life, replacing high-fat whole foods such as nuts, avocados, and cheese. Particularly problematic is the amount of high-fructose corn syrup in soda.

The extreme amount of both sugar and high-fructose corn syrup consumed today is a real problem. The excess glucose is stored as fat if the cells don't need it for energy, but the rapid rate of absorption of fructose in processed foods and soda hits the liver all at once and becomes overloaded. The liver isn't designed to handle that much fructose at one time. Eating more than a small amount of high-fructose corn syrup, white or brown sugar (because it's 50 percent fructose), agave nectar (high in fructose), or refined honey at one time can cause the fructose to spill over into a metabolic pathway, leading to harmful fat production around the middle.[22]

FRUIT ALERT

The good news is that the fructose found in whole-fruit form is absorbed slowly enough that the liver can handle plenty of fruit, even at one time. Fructose isn't inherently toxic. Neither is the glucose that is broken down from carbohydrates and provides energy to the cells in your body. The body handles both of these sugars efficiently when eaten in whole foods, closer to how Mother Nature made them.

SODA AND OTHER SUGAR-FILLED DRINKS

Soda and sweet beverages are the greatest sources of processed sugar in the American diet. The fast rate of absorption of sugar from so many sugary drinks is making this country sick and heavy. Typical table sugar (sucrose) and high-fructose corn syrup (HFCS, found in soda and other processed foods) are made of two molecules called glucose and fructose, in supposedly equal amounts. The main difference between the two is that sugar comes from plant sources like sugar cane, while high-fructose corn syrup is often long chains of glucose from corn. More than half of the glucose is converted into fructose in a lab to make it sweeter. We get too much of both!

The average American drinks about fifty gallons of soda and other sweetened drinks per year. These beverages are the most common way to take in excessive amounts of glucose and fructose. Too much glucose adds to your body-fat stores, and too much fructose overloads your liver and ends up as fat around your middle. This type of belly fat is particularly likely to lead to diseases like diabetes and heart disease. Soda delivers a lot more fructose all at once to the liver.

SODA INVESTIGATION

In 1984, Coca-Cola and Pepsi stopped using sugar in their soda and replaced it with high fructose corn syrup (HFCS). The HFCS used in soft drinks is supposed to contain close to 45 percent glucose and 55 percent fructose. Researchers at the University of Southern California's Keck School of Medicine bought twenty-three popular sodas. They sent the sodas to a laboratory using high-performance liquid chromatography to find out how much sucrose, fructose, and glucose were in each soda. The samples were labeled in code and were tested three times. More than expected fructose levels (which is sweeter) were discovered. The sweeteners in Pepsi and Coca-Cola had as much as 65 percent fructose (and 35 percent glucose), and Sprite contained 64 percent fructose (and 36 percent glucose).[23]

NEW LABEL LANGUAGE

Food labels can be tricky, but the new label laws[24] will help.

Labels are great in theory. But, in practice, you might be surprised by how many loopholes food manufacturers can find when it comes to making their labels say what they want them to say.

Sugar is a great example. When you look at a label, it gives you the total amount of sugar in the product. But what it doesn't specify is how much sugar is naturally found in the food, and how much of it is added after the fact. There is a huge difference between the two.

So you have to dig a little deeper. Go down to the ingredient list and see where sugar falls on the list. Take a look at what types of sugar are included there and in which order. By comparing that with the total amount of sugar in the product, you can get an idea of how much of it is natural and how much of it isn't.

Added sugar exposed

In 2014, the Food and Drug Administration (FDA) announced a long-awaited new labeling system for the food industry that could solve the problem of misleading labels. Most food manufacturers will be required to use the new label by January 2020.

Nutrition Facts	
8 servings per container	
Serving size	**2/3 cup (55g)**

Amount per serving	
Calories	**230**

	% Daily Value*
Total Fat 8g	10%
Saturated Fat 1g	5%
Trans Fat 0g	
Cholesterol 0mg	0%
Sodium 160mg	7%
Total Carbohydrate 37g	13%
Dietary Fiber 4g	14%
Total Sugars 12g	
Includes 10g Added Sugars	20%
Protein 3g	
Vitamin D 2mcg	10%
Calcium 260mg	20%
Iron 8mg	45%
Potassium 235mg	6%

* The % Daily Value (DV) tells you how much a nutrient in a serving of food contributes to a daily diet. 2,000 calories a day is used for general nutrition advice.

Center for Food Safety and Applied Nutrition. "Changes to the Nutrition Facts Label." U.S. Food and Drug Administration, FDA, 18 June 2019, www.fda.gov/food/food-labeling-nutrition/changes-nutrition-facts-label.

The new labels will specify how much sugar is added. This information is particularly helpful in spotting which yogurt is full of added sugar since one cup of plain Greek yogurt already contains 9 grams of sugar naturally. Plain yogurt doesn't have any added sugar, so the new label will indicate that there are 0 grams of added sugar. You can add whole-food sweeteners to the plain yogurt yourself, or at least look for yogurt with less added sugar.

If sugar were reported in teaspoons instead of grams, it would be easier to understand the magnitude of sugar in soda. A 20-ounce bottle of Coca-Cola with 65 grams of sugar has 16 teaspoons of added sugar. Remember there are 4 grams of sugar in every teaspoon so that you can do the math.

Serving size improved

Other improvements include serving sizes that better reflect how much consumers actually eat in one sitting. For instance, an ice cream serving will become two-thirds of a cup rather than the current half-cup. Calorie counts will be in a larger, bolder font so they will be easier to spot.

Remember, most whole foods don't need labels at all. Fruits, vegetables, beans, lentils, whole grains, fish, and nuts are the best bet for your health, whenever you can buy them.

WHOLE AND ANCIENT GRAINS

Fruits and vegetables aren't the only things packed full of nutrients. Whole grains are too[25]—especially ancient grains.

Carbohydrates have gotten a bad rap, and processed grains deserve it. Besides being absorbed very quickly into the bloodstream and putting your body into fat-making gear, refined flour and white rice are stripped of their bran and germ. The bran and germ are loaded with vitamins, minerals, fiber, and thousands of disease-fighting phytonutrients. Whole grains take longer than processed grains to break down. That means that the bloodstream absorbs them more slowly, resulting in more stable blood sugars for better hunger control, long-lasting energy, and more constant fuel for your brain to function at its peak for longer.

Whole grains also feed and increase good bacteria in the large intestine for better digestion, greater absorption of nutrients, and a stronger immune system. It's not surprising that whole grains have been shown to reduce the risk of heart disease, type 2 diabetes, obesity, and even some forms of cancer.

Grains that pack an extra punch

Whole grains are good for you in moderation, but some of them pack an extra punch. Ancient grains are grains that have been around for millennia. While corn, rice, and modern wheat have been bred selectively over time, ancient grains still have a composition similar to what they had thousands of years ago, which can be more beneficial to your body.

Ancient grains include quinoa, farro, sorghum, amaranth, einkorn, millet, spelt, teff, freekeh, and kamut. If you're looking for a good place to start experimenting, farro is one of my favorites. It has a nice, neutral flavor like that of rice.

Farro is bigger in size than other ancient grains, so it feels more satisfying to chew. The pearled version only takes fifteen minutes to cook and still contains 5 grams of fiber and 7 grams of protein in just over half a cup cooked. As a bonus, the wheat-based ancient grains, like farro, tend to be lower in gluten than their modern-day counterparts.

GLUTEN-FREE GRAINS

Consuming wheat has become a much bigger problem for many people in recent years. Whether you've been diagnosed with celiac disease or you have an intolerance to wheat, the side effects are difficult, and wheat, along with barley and rye, needs to be avoided. If you have not been diagnosed with celiac disease, some of the lesser-known parts of wheat may be the problem, according to the latest research. The carbohydrate part of the wheat, called fructans, and amylase-trypsin inhibitors (ATIs), may be your trigger to uncomfortable issues rather than the actual gluten.[26]

Processed food triggers

Food companies add wheat gluten to more than just bread, but also to pasta, snacks, cereals, and crackers, and as a thickener in hundreds of foods to increase shelf life and act as a binder. But the extra gluten is just one of many additives in processed food. Recent research has clarified that excessive amounts of processed foods negatively affect our gut microbe balance and can be a part of the cause of food sensitivities and autoimmune diseases.[27]

Food eaten closer to its whole-food form, with thousands of plant chemicals and without all the artificial add-ins, restores the balance of microbes in our gut to avoid many of these health issues. Genetics certainly plays a role, but know that many with a predisposition to autoimmune diseases go their whole lives without having issues because the environment also plays an important role. Too much overly processed food without enough plant-based whole foods can be that environmental trigger for many.

GLUTEN-FREE WARNING

Gluten-free processed food is still processed food, so don't be lulled into the notion that it's good for you! In fact, the evidence is showing that processed food and sugar are promoting an imbalance in our gut microbes. This imbalance is linked to celiac disease and other autoimmune disease. Grains close to their whole-food form, rather than processed, are the healthiest way to eat within almost any diet restriction but especially with celiac disease and other digestive issues.

Gluten-free whole grains

While changing over to eating primarily foods in their whole form, you can avoid your symptoms. Concentrate on eating ancient grains and especially those void of gluten if you have celiac disease.

Quinoa, brown and black rice, and sorghum are delicious and easy examples of gluten-free whole grain for those needing to eliminate gluten. Gluten-free whole grains (or in some cases seeds or grasses that act like grains) include the following:

- quinoa
- sorghum
- whole-grain corn
- oats (labeled gluten-free)
- popcorn
- brown rice
- wild rice
- black rice
- amaranth
- buckwheat
- chia seeds
- millet
- teff

FODMAP WHAT?

Many people who feel better on a gluten-free diet don't realize that the carbohydrates in the wheat may be causing their intestinal distress rather than gluten. Certain carbohydrates and not just the carbs in wheat are harder to absorb into your body. For some people, this can cause digestive problems.

These particular carbohydrates are referred to by the acronym FODMAP. These letters stand for stands for fermentable, oligosaccharides, disaccharides, monosaccharides, and polyols. These carbohydrates are found in some foods and are not always tolerated in people with irritable bowel syndrome (IBS) or other gut issues. However, for most people, these carbs (like certain types of fruits, wheat, dairy, and legumes, to name a few) are healthy, well-tolerated foods.

Low-FODMAP diet

The low-FODMAP diet[28] eliminates these types of carbohydrates for a short time and then adds them back in one at a time. This can help to identify the types of specific carbohydrates that cause intestinal problems for that individual. These groups of carbohydrates tend to provide important nourishment for your health-promoting gut bacteria. Therefore, it's crucial to find out which groups are not a problem for you. The low-FODMAP diet is medically based and is used and researched all over the world for those with IBS and other intestinal issues. It requires individual guidance by a registered dietitian nutritionist.

If you know which diet you need to be on, you can benefit from having healthier options while still avoiding intestinal discomfort. For instance, if you are on a low-

FODMAP diet, you can usually eat some types of sourdough bread. You can improve your digestive symptoms but can still allow yourself a variety of options that can improve your health.

Whole30 diet

The Whole30 diet is similar to the medically-supervised low-FODMAP diet, but without the critical insight of a dietition to determine which group of foods are triggers for your digestive issues. The Whole30 diet eliminates whole food groups unnecessarily. Eliminating sugar is a good thing, but eliminating all dairy and grains (including whole ancient grains) is often unnecessary and less healthy. In contrast, the low-FODMAP diet still allows yogurt, kefir, and hard cheeses even though most dairy products are excluded for a trial period because they are very low in lactose and made using microbes. Those important details get lost in the oversimplification of the Whole30 plan.

If you want to eliminate a whole category of food, exclude sugar, refined grains (like white flour), and processed food! If whole wheat is an issue for you, eat other whole grains instead. There is a whole grain out there for everyone!

FIX 1

SWAP SUGAR

Replace sugar, refined white flour, and processed foods with whole foods.

If you're ready to lose weight, reduce inflammation, and to feel and think better, the first step is simple: eat less sugar and food made with white, refined flour. When you eat less, you'll want less. Overly processed foods encourage us to crave sugar and refined carbs. The more you avoid sugar and processed food, the less you'll want it over time.

Avoid added sugar and food made with refined flour

Processed food is filled with sugar and high-fructose corn syrup—even when it's not sweet. Prepared salad dressings, boxed mixes, processed syrup, cookies, bread, you name it; if it's not prepared by you or homemade by a friend, it probably has extra sugar or high-fructose corn syrup in it.

Eat sweeteners and food in their whole-food form

Replace refined, processed sugar and high-fructose corn syrup with natural sweeteners such as whole fresh fruit, raw honey, or pure maple syrup. Replace white flour with whole-grain flour and swear off white bread and buns as much as possible! And, of course, make an effort to eat all your food in whole forms, such as whole-grain bread or plain yogurt mixed with a little real maple syrup and topped with fruit and nuts.

Make the swap

This one fix alone can be a game-changer. My clients drop several inches from their waistlines by cutting back on sugar, white bread, white rice, and pasta alone. Even when they're not very strict about it, they still lose weight over the course of several months. Remember, it's never all-or-nothing. If you eat a slice of white bread at a restaurant one day, don't use that as an excuse to go back to eating it all the time.

Action plan

- ✓ Cut out added sugar completely for two weeks to dull your cravings for sweet things. When you need to eat something sweet, eat a piece of fruit in its whole form.
- ✓ Try skipping white, refined bread most of the time. Replace foods made with white flour—such as crackers and bread—with whole grains.
- ✓ Top plain yogurt with fruit and nuts, drizzled with a little pure maple syrup or raw honey.
- ✓ Avoid food products with added sugar. Check the new labels indicating the amount of sugar added.
- ✓ Replace white rice and pasta with brown rice and whole-grain pasta. Or, even better, replace these processed foods with ancient grains such as farro, sorghum, or quinoa.
- ✓ Replace sugary drinks like soda with water or sparkling water.
- ✓ When you need to use sugar in preparing, cooking or baking, use crushed or cooked-down fruit, raw honey, or even a little pure maple syrup (not processed pancake syrup). If you're using a high-powered blender for your recipe, replace sugar with pitted dates.
- ✓ Don't live without chocolate if you want it. Just eat no more than one to two squares of it per day, and make sure it has a 70 percent cocoa content or higher. You can also add cocoa to recipes such as stew, chili, and oatmeal.
- ✓ Order brown rice or other whole grains offered instead of white rice in restaurants.
- ✓ Replace white potatoes with purple and sweet potatoes often.
- ✓ Skip the white bread or bun when eating out by ordering a salad instead of a sandwich. Even skipping half the bun will help.
- ✓ Track your progress on the *12 Fixes to Healthy* app.

FIX 2
PLAN PROTEIN

All 12 Fixes are important, but Fix 2 has the power to help you feel full longer, curb cravings and emotional eating, keep your brain's fuel steady, and help preserve and build your strength as you move more.

People are shocked at the results after eating (not drinking) between 30 to 40 grams of protein for breakfast. Executives are surprised that they are more productive at work before lunch, and so many people are amazed at how their desire to snack at night all but goes away. Give it a try and see what you think.

Fix 2 is to eat a total of 75 to 150 grams of protein spread evenly throughout the day, with at least 30 grams of protein for breakfast. More power to you!

THE POWER OF PROTEIN

Protein is a big power player in creating maximal health. When we eat enough protein over the day, we feel less hungry and more satisfied.[1] Eating protein foods can also help keep fat off and maximize our movements during the day to build muscle. However, the amount of protein, the type, and the times that we eat our protein do matter. While most of us eat enough protein to get by, we often fail to take advantage of the optimal times of day to eat health-promoting, protein-rich foods.

Age-related muscle loss

Around the age of thirty, we begin to lose 3 to 8 percent of our muscle mass every year.[2] Without enough protein spread throughout the day to supply the necessary amino acids, we can't build valuable lean tissue to fight off that muscle loss. And those muscles burn calories for us twenty-four hours a day, seven days a week—even when we're at rest.[3] That helps us to achieve and maintain a healthy weight.

The body doesn't make all the essential amino acids that it needs. Nine of them come from protein sources such as seafood, poultry, meat, eggs, milk and milk products, quinoa, soybeans, and amaranth. While beans, dried peas, lentils, nuts, and peanuts don't contain all nine amino acids, they are protein-rich foods.

The incomplete proteins can easily be paired with another incomplete protein such as rice or corn tortillas to guarantee the intake of all the necessary amino acids. If you eat a variety of foods including vegetables and whole grains, it's easy to take in missing essential amino acids. Or simply eat a complete and healthy protein, such as yogurt or fish, along with them.

Build muscle

Protein helps you to build and maintain muscle mass. Muscle mass is good, compact weight that doesn't take up as much space as fat does while also increasing your strength. Plus, being strong improves your quality of life at any age, but especially as you get older.

The more muscles you have, the more calories you burn—even at rest. Ten pounds of muscle burns about 50 calories a day at rest.[4] Compare that to ten pounds of bulky fat, which burns only 20 calories a day at rest.

A study in the *Journal of Nutrition* in March 2009 compared the effect of a moderately high-protein diet with that of a high-carbohydrate diet over the course of twelve months. Both test groups experienced weight loss. However, the high-protein subjects lost more body fat than the high-carbohydrate dieters did.[5]

Protein spread

You need somewhere between 75 to 150 grams of protein per day, spread out between meals and snacks, to help preserve and increase muscle mass and keep you satisfied. Your body can't store the protein you eat, so it's vital that you spread your protein throughout the day. Additionally, the body can only use a certain amount of protein at a time. If you take in more than 40 to 50 grams in one sitting, the extra protein turns into energy (or fat stores), leaving you without extra muscle benefit.[6]

Time to eat protein

Eat health-promoting protein at every meal and snack, and try to spread your protein throughout the day. If your proportion of protein is out of balance, consume more of it earlier in the day, rather than in the evening, as is typically done in the US. Most Americans eat 60 percent of their protein for dinner.

When you eat during the day, and especially when you eat a high-protein breakfast, you're less hungry during the day and even through the evening. In fact, wholesome breakfast protein makes processed snacks easier to resist.

PROTEIN AND YOUR KIDNEYS

You may have heard that too much protein can harm your kidneys. That's true if you have kidney disease, but this amount of protein is not a problem for healthy kidneys. In fact, 75 to 150 grams of protein is well within the Institute of Medicine's recommended range of 10 to 35 percent of calories from protein. However, this fix recommendation isn't even considered a high-protein eating plan.

Studies in the past have not reported kidney problems with a moderate protein diet.[7] In fact, a 2016 study from the *Journal of the International Society of Sports Nutrition* reported no ill effects on the kidney or liver in healthy adults eating more protein than this recommendation.[8]

High-protein breakfast

Only half of Americans eat breakfast. Yet breakfast really is the most important meal of the day. Eating breakfast enhances weight loss,[9] improves memory, and boosts performance at work or school. But that doesn't mean that all breakfasts are created equal. The best breakfast to sharpen your brain and slim your body is a breakfast rich in healthy protein.

According to Dr. Heather Leidy,[10] lead researcher and assistant professor in the Department of Nutrition and Exercise Physiology at the University of Missouri, protein-rich breakfasts reduce hunger and keep you satisfied all day long. Breakfast that is chewed rather than sipped suppresses the reward-driven part of the brain, so you're less likely to crave less healthy, processed snacks in the late afternoon and night when your willpower is particularly weak. When you eat a high-protein breakfast, you increase your nutrient intake and lower your calorie intake throughout the day. That makes it easier to lose weight!

Dr. Heather Leidy suggests that a breakfast containing between 30 to 40 grams of protein is the best way to achieve fullness and curb food cravings. Brain scans done on participants in Dr. Leidy's many studies[11] have shown that eating more protein

TRY THESE 30-GRAM PROTEIN BREAKFAST IDEAS TO GET STARTED

EGGS:

- 3-**egg** omelet + 1 ounce **cheese** + **vegetables**
- 3 **eggs** + 2 tablespoons **cottage cheese** + 1 ounce **cheddar cheese** + **salt** & **pepper** for scrambled eggs
- 3 **eggs** + 4 tablespoons **cottage cheese** + **salt** & **pepper** for scrambled eggs
- 2 **eggs** + 1 ounce **cheese** + 2 slices **whole-grain toast**
- 2 **eggs** + 1 ounce **cheese** + ½ cup **black beans** + **salsa** + **avocado** + 1 corn **tortilla** (optional)
- 1 **egg** + 1 slice **Canadian bacon** (or 2 eggs instead) + 1 ounce **cheese** + 1 **whole-grain English muffin**

GREEK YOGURT:

- 1 cup **Greek yogurt** + ⅓ cup **granola** + **fruit**
- 1 cup **Greek yogurt** + 2 tablespoons **almond butter** + 2 slices **whole-grain toast**
- 1 cup **Greek yogurt** + 3 high-protein **pancakes** (made with **cottage cheese**) + **peaches**
- ¾ cup **Greek yogurt** + ¼ cup **cottage cheese** + ¼ cup **almonds**

COTTAGE CHEESE:

- 1 cup **cottage cheese** + **fruit** + 1 tablespoon **nuts**
- ¼ cup **cottage cheese** + ¾ cup **Greek yogurt** + ¼ cup **almonds**
- ¼ cup **cottage cheese** + 3 **eggs** + **salt** & **pepper** for scrambled eggs
- 2 tablespoons **cottage cheese** + 3 **eggs** + 1 ounce **cheddar cheese** + **salt** & **pepper** for scrambled eggs
- 3 high-protein **pancakes** (made with **cottage cheese**) + 1 cup **Greek yogurt** + **peaches**

MILK:

- ½ cup **oatmeal** made with ¾ cup **milk** + 2 tablespoons **almonds** + 3 tablespoons **hemp hearts**
- 1 cup **quinoa** cooked with 2 cups **milk** + ½ cup **almonds** + **blueberries** + **cinnamon** + **raw honey**

in the morning decreases activity in the reward-driven, pleasure-seeking part of the brain associated with food cravings later in the day and even at night. Even better, Dr. Leidy and her team have found that high-protein breakfasts increase activity in the executive decision-making part of the brain.

If you can't get at least 30 grams of protein at breakfast, don't give up on this fix! Try adding a mid-morning protein snack paired with a fruit or vegetable to take you the rest of the way.

Keep your focus

Studies have found that eating breakfast may improve your cognitive ability, enhance memory, and increase your attention span.

Because you've been fasting for eight to twelve hours overnight, your brain's fuel source, glucose, is low first thing in the morning. That causes a lack of concentration, alertness, energy, and difficulty recalling information. Eating breakfast in the morning prevents these symptoms by supplying your brain with the fuel it's lacking.

But it matters what you eat in the morning to keep that supply of fuel up and steady. Breakfast should be rich in protein and fiber-rich carbohydrates like oatmeal and certain whole fruits. The carbohydrate is broken down for the actual fuel, while protein and fiber help keep the brain's important fuel level from dropping.

Beyond keeping the brain's fuel source steady, choline is a vital nutrient found in foods like egg yolks and nuts that is important for the creation of memory cells.[12]

A scientific review of thirty-eight studies examined the impact of breakfast on our cognitive abilities.[13] Adults showed a small but robust advantage for memory (particularly delayed recall) from consuming breakfast. Attention, motor, and executive functions were also improved while there were no effects on language.

HOW MUCH PROTEIN?

How much protein should you get, and when should you get it?

According to Donald Layman, PhD, professor emeritus of nutrition at the University of Illinois, we need around 30 grams of protein in one meal[14] to stimulate muscle building, depending on your size (possibly up to 40 grams if you are a large male). However, more than 50 grams of protein per meal is too much and doesn't benefit our muscles.

Aim to take in about 30 grams of high-quality protein each meal for a total of about 90 grams (or at least 75 grams) of protein spread throughout the day. Of course, some of those 90 grams can be included in your snacks to help you boost your intake.

Moderate protein amount

This may seem like a lot of protein, but it is actually considered a moderate protein intake. Those 90 grams of protein amount to only 360 calories—only 18 percent of a typical two-thousand-calorie diet. The Institute of Medicine recommends that you get 10 to 35 percent of your calories from protein,[15] so 90 grams is even at the lower end of that range.

Specific to you

To be more specific to your body size, multiply your current weight in pounds by 0.6,[16] and that will give you an estimated number of grams of protein you need per day to promote body-fat loss while building muscle: 175 lbs. × 0.6 = 105 grams of protein per day.

Use a factor of 0.7 instead of 0.6 if you want greater fat loss, if you're more than sixty-five years old, or if you participate in intense exercise for more than an hour a day: 175 lbs. × 0.7 = 122 grams of protein per day. If you're an athlete in competitive sports, you may need to use a factor of 0.8.

Don't forget that WHEN you consume protein matters too. Spread your total protein intake across all meals and snacks. Don't have a low-protein breakfast, some

> **PROTEIN RDA IS LOW**
>
> The importance of getting enough protein to prevent muscle loss, lose weight, and gain muscle is clear in the research.[17] Yet frequently, the current recommended daily allowance (RDA) for adults (0.36 grams per pound of body weight or 0.8 grams per kilogram) is cited as a reason why we don't need to eat more protein. However, that recommendation is defined as the minimum amount that you need to live, not a healthy optimum. It's not enough protein to offset age-related muscle loss or to help build muscle strength, important for all of us as we age.
>
> Over the past decade, researchers have found that increasing protein consumption is good, safe, and can help prevent muscle loss when losing weight and aging. It certainly does matter when and what type of protein we take in, but those who cite the RDA for protein as a reason to eat less protein are dismissing the updated research.[18]

HOW MUCH PROTEIN ARE YOU EATING?

DAIRY

6 grams:
- 1 string cheese

7 grams:
- 1 egg
- 1 ounce cheddar cheese
- 1 ounce mozzarella

8 grams:
- 1 cup soy milk
- 1 cup milk

11 grams:
- 1 cup yogurt

20 grams:
- 1 cup greek yogurt

28 grams:
- 1 cup cottage cheese

GRAINS

6 grams:
- 1 cup oatmeal

9 grams:
- 1 cup cooked quinoa

11 grams:
- ½ cup buckwheat

13 grams:
- ½ cup amaranth

FISH AND SEAFOOD

17 grams:
- 3 ounces shrimp

19 grams:
- 3 ounces cod

22 grams:
- 3 ounces salmon

NUTS AND SEEDS

6 grams:
- ½ cup walnuts
- ½ cup pumpkin seeds

10 grams:
- 3 tablespoons shelled hemp seeds
- ½ cup almonds

17 grams:
- ½ cup peanuts

MEAT AND POULTRY

9 grams:
- 1 lean sausage link

21 grams:
- 3 ounces lean beef

22 grams:
- 3 ounces pork

26 grams:
- 3 ounces chicken

PULSES

8 grams:
- ½ cup black beans
- 2 tablespoons peanut butter

9 grams:
- ½ cup cooked lentils
- ½ cup white beans

11 grams:
- ½ cup shelled edamame
- 1⅛ cups edamame pods

20 grams:
- ½ cup tofu

34 grams:
- ½ cup dry roasted soybeans

protein at lunch and excessive protein at dinner, as incorrectly done by so many people.

Track your protein

Track your protein the way you track everything else: focus on the fix, but don't be too meticulous and make yourself crazy. Everyone is different, and these numbers are just estimates. Try to eat healthy protein throughout the day but especially in the morning. Include protein sources such as Greek yogurt, eggs, beans, lentils, dried peas, nuts, seeds, or fish at each meal and snack. Use less meat, pork, or poultry.

PROTEIN FOR FITNESS

Your muscles need protein to recover and build. But people still operate on the premise that they need to consume protein within thirty minutes to two hours after they exercise to optimize their gym time. In fact, it may not matter if you consume protein before or after exercising in terms of maximizing muscle growth and repair, as long as you are consuming enough in a twenty-four-hour period and spreading it out.

Newer research indicates that your muscles continue to build for twenty-four hours after a workout. More recent recommendations for protein suggest regularly spacing of moderate amounts of protein throughout the day rather than depending on protein supplements or power bars immediately following a workout.[19]

PROTEIN SHOPPING TOUR

Eat health-promoting protein at every meal and snack, and try to spread your protein throughout the day. If your proportion of protein is out of balance, consume more of it earlier in the day rather than in the evening as is typically done in the US. Also important is eating less of your protein from meat and poultry and more from seafood, plants (such as dried peas, beans, lentils, and nuts), eggs, and yogurt as recommended in the 2015–2020 USDA Dietary Guidelines for Americans, accompanying MyPlate.[20]

Protein update alert!

The newest scientific research indicates that there are a few non-plant foods that improve your health in ways that strictly plant-based diets don't. Eggs from hens that forage for insects, sustainable seafood, and dairy foods made using microbes are health-promoting, non-plant-based foods for which you get a green light.

Protein foods from plants (such as beans, dried peas, lentils, nuts, seeds, and quinoa) are clearly critical for your health. The research indicates that the majority of your food should be plant-based, whole foods! But a few of these healthier non-plant foods can improve your health even more. Vegans are prone to deficiencies in vitamin B12, calcium, iron, zinc, active omega-3 fatty acids EPA and DHA, and fat-soluble vitamins like A and D. These non-plant-based foods prevent these deficiencies while providing quality protein with brain-protecting and gut boosting benefits.

That said, these 12 Fixes are flexible, and any restriction can easily be integrated with this plan. There are plenty of ways to eat a completely plant-based diet using these 12 Fixes. In fact, this protein fix can help ensure that you are getting enough quality plant-based protein by way of whole foods.

Eggs

Eggs are a great source of protein anytime. While egg yolks have gotten a bad rap in recent years, studies have shown that they are really the good guys after all. They are full of important nutrients that people tend to lack in their diets such as choline, biotin, selenium, B12, and vitamin D. The cholesterol in eggs doesn't negatively affect our health like it was believed in the past for most people, and eggs from hens eating insects contain beneficial omega-3 fatty acids as well as more nutrients than commercial eggs.[21]

Buy healthy eggs

Not all eggs are equally nutritious, and it's important to know how to read egg labels to see the difference. Phrases such as "cage-free" and "free-range" on egg cartons can be misleading. Just because chickens aren't in cages doesn't mean they're spending time outdoors in a pasture to increase the nutrient levels of the eggs. They're not. They are packed into a room full of chickens without cages. However, free-range hens do have access to a designated outdoor area.[22]

The term "organic" on the label doesn't guarantee that the hens have a natural lifestyle, either. It means only that the hens are uncaged and have some outdoor access, although how much isn't specified. It also means that the hens are fed an organic, all-vegetarian diet free of antibiotics and pesticides.

HOW DO YOU IDENTIFY THE BEST EGGS?

The healthiest eggs for you are true barnyard eggs, where a chicken is able to freely roam, eating bugs and worms to their heart's content. If you end up in the grocery store buying eggs, look for egg cartons that claim pasture-raised eggs. "Pasture-raised" has no legal meaning so the hens might not be outside scavenging for bugs. Look for information on the label that indicates how much land the hens get to roam on.

If the egg carton is stamped with a "Certified Humane Pasture-Raised" or "Animal Welfare Approved" on the carton, that will ensure their hens are roaming around, pecking away. If the egg carton is just stamped "Certified Humane Raised & Handled" without the word "pasture" included, find some indication on the label that the hens are really outside foraging. Sometimes the label will specifically say that the hens roam in the grass and even give an amount of land per chicken.[23]

Chickens are not vegetarians

It may come as a surprise to you, but despite all the egg carton labels enticing you to buy eggs that are from "vegetarian chickens," hens are not vegetarian. Hens are omnivores; a hen's natural instinct is to forage for bugs and worms.

The nutrition of an egg has a great deal to do with what a hen eats and how much sun the hen is exposed to. Those worms and bugs eaten by the hens boost the egg nutrition by a significant amount![24] In fact, eggs coming from these pasture-roaming chickens have two times as much vitamin E,[25] two thirds more vitamin A, seven times more beta carotene, three times the amount of vitamin D,[26] a third less cholesterol, a quarter less saturated fat, and at least twice the omega-3 fatty acids[27] than regular commercial eggs. Not to mention that eggs from foraging hens also deliver more of the antioxidants lutein and zeaxanthin, both important for eye health.

Given that most consumers perceive that eating eggs from a "vegetarian chicken" is better, many egg producers feed chickens a vegetarian diet. A vegetarian diet does ensure that chickens are not fed animal byproducts (like ground-up chicken) but prevents the hens from going outside to peck for their grub.

Bottom Line: Don't buy eggs that are labeled "vegetarian-fed." Chickens are omnivores after all!

Seafood surprise

Seafood is surprisingly proving to be the real superfood. USDA guidelines recommend eating two to three seafood meals (eight to twelve ounces) per week.[28] The research on seafood's health benefits is very strong, and the mercury issues associated with seafood are now clearer and easier to maneuver.

Mercury toxicity is not an issue in most varieties of seafood, contrary to what was once suspected. Eating fish doesn't cause mercury toxicity; it actually prevents it! The selenium found in most seafood neutralizes the mercury and other toxins, and you often end up with a surplus of selenium reserves for more health benefits.[29]

Dairy protein

Whey protein, found in milk, is easily absorbed and retained by the body. It's also the richest source of leucine—an amino acid that triggers muscle building. Cottage cheese has the most leucine of any dairy food with about 28 grams of protein per cup. Greek and Icelandic yogurts are also concentrated sources of protein with more than double the amount of protein found in traditional yogurt.

A non-dairy diet seems to be all the rage these days. It's seen as healthier, but is it really? Higher protein diets require calcium-rich foods to protect your bones.[30] High-protein, gut-friendly, leucine-rich yogurt or cottage cheese may be just the solution. Sometimes, trends contain some truth but are oversimplified enough that the real truth is hidden.

CULTURED DAIRY NEW FLASH

Of all dairy foods, yogurt, cheese, and other fermented dairy foods offer the greatest benefits for your health! The scientific studies indicate that fermented, cultured dairy products like yogurt and quality cheese have an edge over milk and milk products not made with microbes.[31]

Dr. Dariush Mozaffarian, dean of the Tufts Friedman School of Nutrition Science and Policy, foresees that our nutrition recommendations in the future will focus on different foods rather than specific nutrients. Instead of recommending low-fat or high-fat dairy, for instance, yogurt and cheese (made using microbes) will specifically be encouraged.[32] Dairy foods are not created equal, so if you need to eliminate dairy from your diet, just skip dairy that isn't made with helpful bacteria.

Yogurt power

Yogurt's power to improve your health may be worth including in your diet (even if you don't tolerate dairy foods). Yogurt is an efficient, easy way of taking in enough muscle-enhancing protein during the day. Other needed nutrients, such as calcium, potassium, phosphorus, vitamin D, and B12, come along for the ride. Dairy is linked to improved bone health in children and adolescents[33] and a reduced risk of cardiovascular disease and type 2 diabetes as well as lower blood pressure in adults. Yogurt eaters, in particular, have been found to have lower circulating triglycerides, glucose levels, blood pressure, and insulin resistance than non-yogurt consumers.[34] Don't forget the Harvard study, mentioned in the introduction, found yogurt to have the greatest impact on weight loss, with 120,877 US women and men followed over twenty years.[35]

PROTEIN SHAKE CAUTION

To stay full and satisfied during the day, choose to eat rather than drink your protein. Liquid protein sources are usually digested more quickly than food; and there is plenty of data illustrating greater satiety, reductions in food intake, and weight loss when eating protein food instead of drinking it.

Eating (but not drinking) 30 to 40 grams of protein in the morning tends to suppress the reward-seeking part of the brain, so you have fewer cravings in the late afternoon and at night when you're more vulnerable.

Some people do benefit greatly from drinking a protein shake, however. Liquid protein is especially helpful for the elderly, those participating in extra-long workouts, and cancer patients who struggle to consume enough food.

Dairy made with microbes

Some are opposed to dairy, but the truth is that the powerful probiotics in fermented dairy products, such as yogurt and kefir, may counter and even surpass the perceived health negatives of milk products.[36] Of course, some people are allergic to the protein in milk and shouldn't add it back to their diet. However, for those who are lactose intolerant or who have a sensitivity to dairy, fermented dairy may be a viable option and well worth a try. Yogurt is low in lactose because the good bacteria in it breaks lactose down during fermentation. Kefir is even better; the bacteria in it are so strong that it's actually 99 percent lactose-free.

Cheese surprise

Quality cheese is often created, in part, by adding live bacteria cultures, fermenting the milk's natural sugars into anti-inflammatory lactic acid. Romano (Pecorino Romano), Gouda, some cheddars, Parmesan cheese (Parmigiano-Reggiano), and even cottage cheese are made using this method. Many of these cheeses still contain some live bacteria. These cheeses are very different than processed cheese like American cheese slices.

FIX 2

PLAN PROTEIN

Eat a total of 75 to 150 grams of protein spread evenly throughout the day, with at least 30 grams of protein for breakfast.

Consuming enough protein in your diet will be a powerful weapon in an attempt to fight obesity and preserve your muscle strength.

Protein amount for you

Multiply your current weight in pounds by 0.6 for an estimated number of grams of protein per day to improve body composition: 175 lbs. × 0.6 = 105 grams of protein per day.

To promote more body-fat loss while building muscle, you can multiply your weight by 0.7 instead of 0.6. Also, those in their late sixties or older should multiply by 0.7 to offset age-related muscle loss. Using a factor of 0.7 can also have advantages for some athletes and for certain days filled with extended, intense exercise.

Protein to eat less

Protein has a greater satiety effect than do either carbohydrates or fats, so you stay satisfied for a longer period of time. You become better able to control your appetite, and you eat less overall. It also takes more energy to digest, absorb, and metabolize protein than it does to process carbohydrates and fats. The more energy you use, the more calories you burn.

Spread protein out

Eat your protein intake fairly evenly over the course of the day to ensure that protein's muscle-building blocks, amino acids, are always available to build new muscle.

Healthiest high-protein foods

Meals that include nutrient-rich foods from several protein sources are optimal. Protein foods from plants (such as beans, lentils, nuts, seeds, edamame, and quinoa), yogurt, seafood, high-quality cheese, cottage cheese (the highest source of leucine for muscles), and barnyard eggs are the healthiest high-protein sources.

Protein for breakfast

A protein-rich breakfast reduces hunger and keeps you satisfied throughout the day. Studies from the University of Missouri have shown that eating protein in the morning tones down the pleasure-seeking part of the brain associated with food cravings and cocaine addiction. The result? You end up with significantly more control over what you eat—especially late at night when your resistance to indulgence is at its lowest.

Try to eat at least 30 grams of protein for breakfast and between 75 to 150 grams of protein spread throughout the day. I've noticed that when I eat closer to 40 grams of protein for breakfast, I'm surprised by how uninterested I am in eating my pleasure foods in the evening. A high-protein breakfast helps you take in more nutrients all day long, giving you more satisfaction with healthier food and fewer cravings for junk foods. This ultimately results in the loss of body fat.

Protein shakes

Keep in mind that drinking your protein for breakfast doesn't work the same way that eating it does for satisfaction and food cravings, though protein drinks can help build muscle just as effectively. The lack of chewing brings less satisfaction and doesn't deactivate the pleasure-seeking part of the brain like actually eating enough morning protein does.

Track protein

To get the hang of eating enough healthy sources of protein at breakfast and throughout the day, track it for a while. Estimate the amount of protein you eat and track this fix with the *12 Fixes to Healthy* app. After a few days, you'll be able to recall how much protein different foods contain without looking it up. Remember not to stress out about it. It's better to eat around the right amount of protein than to get frustrated and abandon the fix altogether.

Action plan

- ✓ Start your day with 30 to 40 grams of protein for breakfast or between your breakfast and morning snack.

- ✓ Spread out your protein throughout the day, but if you do end up having more protein at one end of the day, make it morning rather than the evening.

- ✓ Include a high-protein food, such as nuts, nut butters, or Greek yogurt, in every snack and meal you eat.

- ✓ Multiply your weight in pounds by 0.6 to see how many grams of protein you should aim to eat every day. Multiply by 0.7 if you are in your late sixties or older or for days filled with extended, intense exercise.

- ✓ Choose from the healthiest higher-protein foods, such as plant-based protein (beans, lentils, dried peas, edamame, nuts, peanuts), low-sugar yogurt, seafood, and barnyard eggs, to eat throughout the day.

- ✓ Add up the grams of protein you eat every day until you get an idea of which foods you need to eat to consume enough quality protein spread throughout the day.

- ✓ Track your progress on the *12 Fixes to Healthy* app.

FIX 3
BOOST GUT

Like wealthy tourists that spend money and help the local economy, live active bacteria found in some foods boost your gut health. Those touristy bacteria don't stick around for long, so you need to bring them in on a daily basis. They can make a difference to your weight, health, and even your brain's reactions to the world around you (anger, anxiety, etc).

So invite those "good" bacteria to visit your gut by following Fix 3 to eat a fermented food every day or at least several times per week. You don't need much, but consistently eating cultured foods with live, active bacteria helps boost the health of your gut and, thus, your overall health.

GUT MICROBES

Three pounds of organisms inside the digestive system include around some forty trillion bacteria, fungi, and viruses and are collectively known as your microbiota. In the past decade, researchers have discovered that the microbiota aren't squatters mooching off your nutrient-rich environment. In fact, this microbiota, often referred to as the microbiome, is now considered a living organ, working with your body to absorb nutrients from food, squeeze out germy invaders, and calibrate and boost the immune system. An imbalance of microbiota in your gut can negatively affect your weight, ability to fight off sickness, risk of disease, and even responses to the world around us.[1]

Disease and gut microbes

The gut microbiome is the balance of these microbes in our gut. Already, recent technology in gut microbiome research shows that subtle imbalances in the gut microbiome affect almost all diseases.[2] The role of the microbiome is evident in the progress and development of diseases like obesity,[3] hypertension, cardiovascular disease, diabetes,[4] cancer, celiac disease, rheumatoid arthritis, multiple sclerosis, inflammatory bowel disease, gout, autoimmune disease,[5] arthritis, depression, autism, mood disorders, infant health, as well as longevity.

For many of these diseases (like diabetes, for instance) there appears to be a different gut microbiome than for those without the disease. It seems simple, right? Well, it's not. Somehow, the microbiome is mysteriously intertwined with the immune system, boosting or depressing health.[6]

The mucosal linings that shield our organs from microbial invaders are part of the system suppressing or destroying harmful bacteria. The microbiome trains the immune system, while the immune system influences the microbiome. It's complicated, but the focus is how to adapt the microbiome to shift the direction of disease.

Gut microbes affect weight

A healthy microbiota seems to help keep our weight under control. The beneficial bacteria appear to lower intestinal inflammation and aid digestion, preventing the buildup of body fat. Several studies have reported that the microbial populations in the gut are different between heavy and lean people and that when heavy people lose weight, their microflora becomes similar to that of lean people.[7] In fact, when the gut bacteria of pairs of lean and obese twins were transplanted into mice, the mice that got the lean twins' bacteria remained lean, while the mice that got the obese twins' bacteria became fatter, even though they ate the same amount.[8]

Gut bacteria affect calories

The research suggests that our ability to extract and store calories from food as fat is at least partially influenced by our gut microbes.[9] In turn, our dietary choices affect our gut microbiome. Do scientists need to redefine a calorie in terms of how the gut microbiome breaks it down?

A study from Purdue University evaluated and compared the fat absorption of ingested peanut butter and whole peanuts. All the fat in the peanut butter was absorbed, while only 62 percent of the fat in whole peanuts was absorbed.[10]

There is no doubt that eating nuts in their whole food form allows many mechanisms to work together to promote a healthy body composition and improve your health. In fact, there is even indication that those who eat nuts regularly in modest amounts tend to live longer.

Gut microbiota help to break down the food you eat and appear to contribute to the whole-food calorie puzzle. They play a major role in digestion, storage, and burning calories in your food. Gut bacteria influence the number of calories absorbed from your diet and may add about 140 to 180 kcal/day to energy absorption.[11]

It appears that low-fiber diets are associated with the microbiome, absorbing more calories out of food and leading to significant weight changes over time.[12] Diets with more plant-based starch and dietary fibers, instead of animal fat and protein, result in more calorie loss in stools, which can be helpful in managing your weight.

It's amazing that gut microbes have so much influence on how many calories you extract from your food. But it's critical to know that food patterns similar to the Mediterranean-style diet (olive oil, fruits, vegetables, nuts, beans, Greek yogurt, and fish) and this *12 Fixes to Healthy* plan have a significant impact on creating a healthy gut community to help burn those calories.

Fruits, vegetables, whole grains, exercise, and sleep were important among participants who lost weight and maintained healthy weight levels in a large Harvard study. This study, published in 2011 in the *New England Journal of Medicine* evaluated the habits affecting the body weight of over 120,000 people over twenty years. Most people in the study gained weight over the years, but some lost weight. Those who ate yogurt lost the most weight: an average of 0.82 pounds every four years, which the researchers attributed to the yogurt's healthy bacteria.[13]

KETO DIET ON YOUR MICROBIOME

Following a keto diet or any type of low-carb diet may help you lose weight in the short term, but your gut balance and diversity of microbes may be left primed for weight gain. Restricting plant-based carbohydrates like fruits, vegetables, whole grains, beans, and lentils can have a huge negative impact on the gut microbiota.

Fiber from whole foods like fruits and whole grains higher in carbohydrates is essential for bacterial diversity. Less fiber can also add to constipation, which is a common side effect of low-carb diets, especially the keto diet. Poor gut bacteria diversity leads to weight gain.[14]

Brain gut connection

When you think of neurons, you probably think of the neurons in your brain. Your digestive system has been dubbed the "second brain," thanks to the millions of neurons that also line your gut. These gut neurons release critical messengers known as neurotransmitters to communicate with the brain and influence your emotions and mood.[15]

Our gut microbiome plays a crucial role in the management of many neurotransmitters that regulate your mood. For instance, about 95 percent of serotonin and 50 percent of the dopamine is made in the gut and makes you feel happy.

The impact of these gut bacteria on your brain function is becoming more apparent as the research is expanding and indicating that our gut microbe balance affects our mental responses.[16] In one study, a combination of the bacteria lactobacillus and Bifidobacterium in yogurt was given to healthy adults for a month. There were significant improvements in depression, anger, and anxiety. Findings from UCLA studies have indicated that bacteria in yogurt may actually change the way our brains respond to the environment, including cognitive function.[17]

Gut health link to celiac disease

Researchers are just starting to piece together the causes for the alarming rise in celiac disease in the last two decades. While there is a genetic component to autoimmune diseases, there is also an environmental component that can activate those genes. The gut microbiota appears to be a part of the equation. Because so many environmental factors known to influence microbes in the microbiome have been directly linked to celiac disease, some studies suggest that microbiota balance is a significant factor in the development of celiac disease.[18]

LIFESTYLE TRUMPS GENETICS

Extensive research with twins reveals that our environment, such as diet and who you live with, is far more significant than the genetic factors in influencing the function and composition of gut microbiota.[19]

IMPROVE GUT MICROBE BALANCE

Hippocrates' well-known quote, "Let food be thy medicine and medicine be thy food," appears to be timelessly wise! Diet plays a powerful role in shaping the balance and diversity of microorganisms in your gut to promote either health or disease.

Like fingerprints, we all have a different microbiome, which is influenced by so much, including delivery at birth, breastfeeding, medications (especially antibiotics), stress, and diet. But the most common negative factor to the microbe balance and diversity in your gut is a food pattern low in fiber from fruits, vegetables, beans, lentils, whole grains, nuts, and seeds but high in processed sugar, animal protein (like red meat), packaged food products, and fried foods.

Dietary changes

Commit to eating a cultured food every day or at least several times per week to add natural bacteria (also called probiotics) to your gut microbiome. While the bacteria in cultured food boost your gut health, they don't stay around for long as they ride out with the stool. Consequently, these beneficial microbes need to be consumed in food on a daily basis. So eat a cultured (also referred to as fermented) food daily.

Whole plant-based foods

Beyond the addition of probiotic foods, add whole plant foods like a wide range of vegetables, fruits, nuts, seeds, and whole grains to boost the development and maintenance of a healthy balance of microbes in your gut. Skip the sugar and processed foods (even if they are gluten-free products) that tend to promote a poor balance of health-promoting microbes in your gut.

PROCESSED GLUTEN-FREE FOOD PRODUCTS

While it's true that the processed, gluten-free products stacked across the shelves at the supermarkets may help a celiac person avoid acute symptoms, these products, like all highly processed foods, negatively affect the gut microbiome, increasing inflammation and propelling disease forward.

To determine if a product is highly processed check the ingredient list. If the food is made up of a significant amount of refined sugar or refined flour or includes ingredients that you can't buy to use in your own kitchen, the food is considered overly processed and should be eaten sparingly. Whole foods in their whole food form (like fruit) and minimally processed foods (cooked quinoa or a collection of unrefined foods cooked together) encourage diversity and enough friendly microbes to help prevent the overgrowth of the bad bacteria.

It doesn't take long to strengthen the microbiota balance with whole foods that feed and nourish the helpful microbes in your gut. In one study it only took two weeks for 248 participants to improve their microbiome with a diet high in vegetables, fruits, and other high-fiber, plant-based foods. Those who ate more fruits, vegetables, and whole grains boosted their gut bacterial diversity and improved blood-sugar levels and metabolism.

For optimal gut health, consume foods that contain probiotics naturally as well as whole plant-based foods on a regular basis. In fact, all the 12 Fixes in this book support and improve your microbiome!

Fiber gut alert

The value of whole foods full of fiber to feed the friendly bacteria in the gut is often overlooked! But your gut bacteria live off whatever's leftover in your colon after your cells have digested all the amino acids and nutrients. Eating a wide variety of foods with plenty of fiber-rich fruits, vegetables, and whole grains is the best way to encourage a balanced and diverse microbiome.[20]

A sudden swap to high fiber foods may cause bloating, so add them progressively and notice how your body reacts. Sometimes you can spot specific foods that make you feel bad when you eat them, and you can swap them for different foods that feel better for you but are still high in fiber and nutrients and low in refined ingredients and saturated fats.

HOW MUCH FIBER?

Limiting dairy (except for cultured dairy), red and processed meats, and refined sugars can improve gut health. But replace these foods with foods high in fiber. The recommended amount of fiber is at least 25–29 grams of fiber from foods per day, yet typically, people eat fewer than 20 grams of fiber per day. I'm not talking about a fiber bar, necessarily, that takes fiber from its original source to boost the processed bar's fiber amount. No. I'm encouraging you to increase your fiber amount by adding whole grains, fruits, vegetables, nuts, and seeds to your diet for the best results to your gut balance and diversity of those little bugs.

Breast milk boosts gut health

Breast milk is the best ally to induce a healthy balance of bacteria in the gut early on. Establishing a healthy microbiome right from the beginning is critical, and

FIBER IN WHOLE FOODS

FRUITS

1.0 gram:
- 1 cup avocado

3.0 grams:
- 1 medium banana
- 1 medium orange
- 1 cup strawberries

4.5 grams:
- 1 medium apple, with skin

5.5 grams:
- 1 medium pear

8.0 grams:
- 1 cup raspberries

VEGETABLES

1.5 grams:
- 1 medium carrot

2.0 grams:
- 1 cup chopped cauliflower

3.5 grams:
- 1 cup sweet corn

3.8 grams:
- 1 medium sweet potato, no skin
- 1 cup beets

4.0 grams:
- 1 medium potato, with skin
- 1 cup brussels sprouts

5.0 grams:
- 1 cup turnip greens
- 1 cup chopped broccoli

9.0 grams:
- 1 cup green peas

10.3 grams:
- 1 artichoke

GRAINS

2.0 grams:
- 1 slice bread, whole wheat

3.5 grams:
- 1 cup brown rice, cooked
- 3 cups popcorn, air-popped

5.0 grams:
- 1 cup oatmeal, cooked
- 1 cup quinoa, cooked

5.5 grams:
- ¾ cup bran flakes

6.0 grams:
- 1 cup barley, pearled, cooked
- 1 cup spaghetti, whole wheat, cooked

LEGUMES, NUTS, AND SEEDS

3.0 grams:
- 1 ounce sunflower seeds
- 1 ounce (49 nuts) pistachios

3.5 grams:
- 1 ounce (23 nuts) almonds

10.0 grams:
- 1 ounce chia seeds
- 1 cup baked beans, cooked

12.0 grams:
- 1 cup edamame, shelled

12.5 grams:
- 1 cup chickpeas, cooked

15.0 grams:
- 1 cup black beans, cooked

15.5 grams:
- 1 cup lentils, cooked

16.0 grams:
- 1 cup split peas, cooked

*Rounded to nearest 0.5 gram.
USDA National Nutrient Database for Standard Reference

breastfeeding fosters beneficial gut bacteria. Because intestinal microbiota affects the development of the immune system, the role of the friendly bacteria influencing the immune system is particularly important in the first few months of life.[21]

Probiotic supplements

What about probiotic supplements? It seems easier to pop a pill than to change your diet. Well, think again, if you want benefits tailored to your body's needs. Every person's gut microbe collection is different and lives in a delicate balance. So, a probiotic supplement that helps one person might not help someone else. Eating fiber in fruits, vegetables, and whole grains gives your body an opportunity to make the microbes specific to your specific needs.

Of all the thousands of probiotic strains, the ones in those supplements may not be the that your body needs. Taking a probiotic supplement is like trying to find a needle in a haystack, unless your doctor is prescribing a specific probiotic supplement for a specific issue.

There are some probiotics identified that help with specific issues. For instance, Align brand is a probiotic that may improve gastric reflux and has been shown to decrease symptoms like bloating, stool frequency, and cramping in those dealing with irritable bowel syndrome (IBS). However, all but a few brands of probiotics haven't been shown to help specific issues. Also, you should know that probiotics are not well-regulated since they are not a drug or a food product. That means you may not be getting exactly what the label claims, and the quality may vary. Given that probiotics are live bacteria, their conditions need to be right too. You may be wasting your money on a probiotic supplement when eating a cultured food

WATER KEFIR WINS OVER KOMBUCHA

Despite kombucha's (fermented tea) popularity for improving gut health, its probiotic strength is fairly weak. Kefir (both water and milk based), on the other hand, is a much better source of friendly bacteria. In fact, kefir grains can have up to fifty-six different yeast and bacterial strains, leaving kefir more potent to encourage a healthier gut environment compared to kombucha.

Unlike kombucha, kefir doesn't contain caffeine and has negligible alcohol content, with less risk of contamination than kombucha. Water kefir wins over kombucha as a substitute for alcohol and soda. Still, there is some sugar in water kefir and kombucha, so don't abandon good old water as your main beverage choice!

containing live bacteria from the fermentation process is more beneficial. In this stage of probiotic research, we don't know enough to outwit nature, so we may just be outwitting ourselves!

Cultured or fermented food?

In this case, *cultured* does not have anything to do with how refined your manners are. Cultured food means that in the process of making, that food has been fermented. Fermented food sounds kind of creepy until you learn what it actually is. It's actually one of the oldest ways to preserve food. During a fermentation process, microbes or good bacteria eat or break down the carbohydrate part of the food to produce lactic acid.[22] Carbs are replaced with lactic acid. That's why sourdough bread, for instance, doesn't raise your blood sugar level as high as regular bread does (not made with microbes). The sour taste is a result of the lactic acid made by the microbes after eating the carbs in the flour.

The lactic acid in fermented food helps your body to digest, absorb, and assimilate nutrients. In some cases, the fermentation process actually creates nutrients for the body and boosts your immune system. This process promotes the growth of a healthy bacterial balance throughout your digestive tract.

It seems like every country has some type of fermented food that has been part of their culture. Fermented soybeans are used in Japan to make miso or tempeh, and there is a traditional fermented vegetable and spice dish in Korea called kimchi. These traditional foods have been shown to be beneficial to your gut health.

Fermented foods are easier to digest

Friendly microbes feed on the carbohydrate of the grains to produce lactic acid in another fermentation process. Whole-grain sourdough bread is a good example. A starter full of good microbes consumes the carbohydrate portion of the flour to rise the bread. The result is fewer carbohydrates and more of a sour-tasting bread. This process naturally neutralizes those plant chemicals (like lectins and phytic acid) that prevent the absorption of nutrients. The nutrient content of your grains is greater, and they are easier to digest[23] and provide better blood-sugar control as well.

Dead or alive

Even though cultured or fermented food is made with microbes, that doesn't mean the good bacteria are alive when you eat it. But if the bacteria are alive in the food, it's often referred to as a probiotic. Many foods are made using live microbes (meaning it's been fermented), but the bacteria used to make the food dies with heat.

Sourdough bread

The microbes used to make sourdough bread are a great example. Sourdough bread is devoid of live, active bacteria by the time you eat it. You won't get a gut boost from the bacteria, but the food is left healthier than if microbes weren't used at all.

Cottage cheese

Cottage cheese is another example of food made with a fermentation process that typically ends up without live microbes. Cottage cheese is pasteurized, and the heat kills the bacteria. It is still left with some perks, however. Cottage cheese is low in the natural milk sugar called lactose.

Cottage cheese has other benefits too. It's full of easily digested whey protein that is efficiently absorbed and retained by the body. Cottage cheese is also the richest source of leucine—an amino acid that triggers muscle building. In fact, cottage cheese has the most leucine of any dairy food and contains about 28 grams of protein per cup along with valuable minerals like calcium.

Now that live microbes are becoming valued, you can often find a cottage cheese that contains live bacteria also. The cottage cheese with live bacteria will be sourer than cottage cheese that does not contain live bacteria, but those live bacteria will benefit your gut health.

Probiotic foods

Probiotic foods are foods that still contain live, active bacteria. The bacteria is positive to your gut microbe balance, but those microbes don't stick around. They get escorted out with your poop, so it's important to eat foods with live bacteria consistently.

Yogurt

Yogurt and kefir are foods that are made using a fermented process and still usually contain live, active bacteria when consumed. The live bacteria is added after the milk is heated in the process of pasteurization. Pasteurization actually kills some of the bad bacteria that can sometimes infect foods and make you sick. You're getting the best of both worlds with yogurt: live, friendly bacteria without any harmful bacteria. Sweeten it with fruit and a bit of raw honey instead of processed sugar for the best impact on your gut microbe balance.

Yogurt is an efficient, easy way of taking in enough muscle-enhancing protein during the day. Nutrients, such as calcium, potassium, phosphorus, vitamin D, and B12, come along for the ride. Dairy is linked to improved bone health in children and adolescents and a reduced risk of cardiovascular disease and type 2 diabetes as well as lower blood pressure in adults. Yogurt eaters, in particular, have been found to have lower circulating triglycerides, glucose levels, blood pressure, and insulin resistance than non-yogurt consumers.[24]

Cheese

Quality cheese is often created in part by adding live bacteria cultures, which ferments the milk's natural sugars into anti-inflammatory lactic acid. Romano (Pecorino Romano), Gouda, some cheddars, Parmesan cheese (Parmigiano-Reggiano), and many different aged types of cheese are made using this method and often contain some live bacteria. The French's good health results are in part due to the health-promoting microbes in the aged cheeses they've been eating for generations.

> **DAIRY FOOD ALERT**
>
> Of all dairy foods, yogurt, cheese, and other fermented dairy foods made using microbes offer the greatest benefits for your health! The scientific studies indicate that fermented, cultured dairy products like yogurt and cheese have an edge over milk and milk products not made with microbes. Dairy foods are not all created equally. Future nutrition recommendations will focus on yogurt and cheese (made using microbes) over other dairy foods. So when someone tells you to eliminate all dairy from your diet, be sure to keep cultured, fermented dairy foods in your diet if possible!

PREBIOTICS

While probiotic foods contain live bacteria, prebiotic foods are fiber parts of plants that we can't digest, so they become food for the good bacteria already living in our digestive systems. Studies report that eating more prebiotics positively affects your health by supporting your microbiota. By feeding the good bacteria, prebiotic foods can reduce your appetite, lower your body fat, improve your glucose tolerance, and improve your mood.[25]

Include a variety of prebiotics in your diet to provide food for different strains of healthy bacteria. Prebiotics are found in fruits, vegetables, legumes, nuts and seeds, and whole grains. Whole wheat, whole barley, oatmeal, avocados, asparagus, Jerusalem artichokes, less ripe bananas, soybeans, legumes, jicama, chicory root, dandelion greens, okra, radishes, onions, garlic, leeks, chives, scallions, raw honey, pure maple syrup,[26] and apple cider vinegar are foods that have specifically been identified as beneficial to nourish the health-promoting microbes in our gut. In general, plant-based foods eaten in their whole-food form appear to help feed disease-fighting gut bacteria and are beneficial to your microbiome balance.

Each type of prebiotic promotes the growth of different kinds of helpful bacteria, so a diet that has a variety of fiber sources supports microbiome diversity. Whole plant foods contain polyphenols. These phytochemicals are nutrients in fruits and vegetables that can slow the growth of toxic bacteria. This allows good bacteria to grow and helps prevent harmful microbes from taking over.[27] Another reason to eat more produce!

GUT RUT

Did you know that a diet consisting of lots of different types of whole foods can lead to a healthier, more diverse gut microbiome? In fact, adding variety to your diet can change your gut bacteria profile in just a few days.[28] But our Western diet is not very diverse. In fact, 75 percent of the world's food comes from just five animal species and twelve different plants. If your diet is in a rut, so is your gut. Eat a variety of different fruits, vegetables including beans and lentils, and whole grains!

Resistant starch

Leftovers, eaten cold or reheated, feed you and the friendly bacteria in your gut! Cooked potatoes, pasta, or rice that has been cooked, cooled, and then reheated (or simply eaten cold after cooking) are full of resistant starch, beneficial to your gut. Resistant starch feeds the helpful bacteria and encourages these good guys to grow and keeps the bad bacteria in check. As resistant starch acts as a prebiotic, it helps your insulin to be more effective at removing sugar from your blood.

Want to make your leftovers a double whammy to boost your gut health? Make leftovers with whole grains or purple or sweet potatoes. The fiber in whole grains like brown, red, and black rice or pasta made with whole grains is great food for the friendly microbes in your gut. Purple potatoes and sweet potatoes have other advantages over white potatoes too.

WHAT YOU EAT MATTERS TO YOUR GUT!

The food we eat primarily determines our gut microbiome! Not only does the good bacteria in some foods like yogurt, kefir, artisan cheeses, kimchi, and sauerkraut boost your gut bacteria balance, but foods like produce, legumes, and whole grains feed the helpful bacteria in your gut. Sugar and refined, processed foods feed the bad bacteria.

Replace sugar, refined flour, and processed food that feed the harmful bacteria with fruits, vegetables, and whole grains that feed the good bacteria.

SUPERMARKET YOGURT SECTION

Yogurt is a quick and nutritious food—but only if you buy the right kinds. Greek yogurt and Icelandic skyr both give you the biggest protein and calcium bang for your health.

Greek yogurt

Greek yogurt is made by straining the extra liquid or whey from regular yogurt. The yogurt becomes thicker, tangier, and creamier than regular yogurt. Also, plain Greek yogurt has more protein and less sugar than regular yogurt. More milk goes into making each cup of Greek yogurt. However, regular yogurt contains more calcium than Greek yogurt. One 5.3-oz. container of plain Greek yogurt contains 90 calories, 15 grams of protein, and 4 grams of natural sugar (not added).

Icelandic yogurt

Skyr is a style of yogurt from Iceland and is pronounced "skeer." Traditionally made from nonfat milk, skyr is a strained yogurt similar to Greek yogurt, except even thicker and with a hint of natural sweetness. The probiotic cultures found in the skyr are also similar to those found in Greek yogurt. Siggi's and Icelandic Provisions are two popular brands of skyr. A 5.3-oz. container of plain fat-free Siggi's contains 80 calories, 15 grams of protein, and 4 grams of natural sugar (not added).

Problems with dairy?

Those who don't tolerate dairy well may be having a difficult time digesting the milk or lactose found in dairy products. Yogurt is very low in lactose because the microbes consume the lactose during the fermenting process. Even the medically developed low-FODMAP diet for digestive problems allows for yogurt, kefir, and hard cheeses like Romano cheese, while most other dairy products are excluded during a trial time.

Dairy products may also be problematic for some people due to the feed that cows are eating these days. Don't completely give up on yogurt from dairy until you've given grass-fed yogurt a chance. If you're not allergic to milk, don't miss out on these healthful ways to take in nutrient-loaded protein with gut-restoring probiotics in whole-food form. The nutrients and protein that dairy offers are hard to beat.

Goat-milk and sheep-milk yogurt and cheese

Yogurt and cheese made from goat's milk is another great option. Give them a try as well. Romano and other cheeses made from sheep's milk and goat's milk tend to be easier to digest. Both goat's and sheep's milk have a greater number of short- and medium-chain fatty acids.

Dairy-free yogurt

Yogurt can also be completely dairy-free when made with nondairy kinds of milk such as soy, almond, rice, or coconut milk. Most of these nondairy yogurts won't provide nearly as much protein or calcium as cow's milk, but they are fermented foods that can improve your gut flora and be helpful to your health.

Regardless of what ingredients the yogurt is made from, many on the market have been corrupted with sugar and artificial ingredients. Follow these seven simple tips to buy the best yogurt for yourself and your family.

The yogurt rules: seven tips to buy great yogurt

1. **Look for yogurt that contains "live and active cultures."**

    Some companies heat-treat yogurt after culturing it with the good bacteria, killing both the good and the bad bacteria to make it last longer and reduce tartness. These live and active bacteria promote gut health, boost immunity, and may even be a key to becoming slim and trim.

2. **Choose yogurt that is low in sugar.**

    Avoid any yogurt that lists sugar as the first ingredient. A standard 5.3-oz. individual container of yogurt should contain no more than 16 grams of sugar and, ideally, fewer than 13 grams. Shoot for fewer than 25 grams of sugar per cup, but fewer than 20 grams of sugar per cup is better. The best option is to add fruit and even a little pure maple syrup or raw honey to plain yogurt. Another good option is to mix plain yogurt or cottage cheese into sweetened yogurt. This dilutes the sugar and boosts the protein all in one step.

3. **Buy yogurt containing real fruit.**

    Some so-called fruit yogurt is really just a mix of sugar and food coloring. Make sure you see actual fruit on the list of ingredients when you buy fruit yogurt—ideally in the list before any added sugars.

4. **Avoid artificial sugar substitutes.**

    "Low sugar" varieties of yogurt are often full of artificial sweeteners. If you need to buy yogurt with an artificial sweetener in it, choose one sweetened with monk fruit. Avoid aspartame and sucralose.

5. **Avoid thickeners.**

    Some brands of Greek yogurt cut costs by adding thickeners such as gelatin, cornstarch, gums, lecithin, carboxymethylcellulose, polysorbate-80, and pectin instead of straining the yogurt to make it thick and creamy (which more than doubles the protein and lowers the lactose). These Greek yogurt impostors are often labeled "Greek style." Authentic Greek yogurt doesn't contain these thickening agents or has minimal amounts of them. A good Greek yogurt should have 12 to 16 grams of protein per 5.3-oz. serving or 19 to 23 grams of protein per cup.

6. **Keep it simple.**

    All you need to make yogurt is milk and live, active bacteria. Plain yogurt should contain nothing more than those two things. Flavored Greek yogurt can have more ingredients, but the list should be short and filled with words you can identify. Especially avoid added hydrogenated fat, artificial sweeteners, or artificial colorings and flavors.

7. **Don't assume that "nonfat" means "healthy."**

    "Nonfat" doesn't always mean "low in calories." Many nonfat yogurts contain a large amount of added sugar. Go for a version that gets most of its sweetness from real fruit. A higher percentage of fat in your yogurt may be more satisfying and appears to be neutral for your health. But because the protein content often goes down when the fat goes up, lower-fat yogurt can be advantageous.

YOGURT PREP

It's yogurt time in the kitchen! Buy yogurt or make it. If you make it, it takes only two ingredients, and you can do it right in a mason jar. Try it at least once to see how incredibly easy, economical, and delicious it is!

Savory yogurt

Throw away the notion that yogurt is just for yogurt parfaits and smoothies. Great as either savory or sweet, yogurt and kefir can be used in all sorts of ways to make food taste better and more nutritious! Both India and the Middle East have delicious, mild versions of cucumber yogurt that are wonderful on all sorts of foods, from spicy curries to chicken. Chicken marinated with yogurt remains juicy even if it is a little overcooked.

Greek yogurt replaces mayo and keeps food creamy and tasty when you add a vinaigrette to match the consistency you desire. Try it in pasta, slaw, and chicken and tuna salads. I even add plain Greek yogurt to mashed potatoes to cream them up and add nutrition. Dips prepared with yogurt instead of mayo often taste better. Mix mashed avocados with yogurt for a delicious dip. Creamy salad dressings made with yogurt are outstanding and so much healthier. You can thicken or top with plain Greek yogurt instead of sour cream for soups, chili, salmon cakes, and so on.

On the sweeter side

On the sweeter side, vanilla yogurt (or add pure maple syrup to plain yogurt), along with fruit, makes a wonderful topping for pancakes, waffles, and crepes. I love to bake with yogurt and kefir to add nutrition while cutting the butter. Plus, when kefir replaces buttermilk in baked goods such as muffins, quick breads, and pancakes, the consistency is silky and smooth. Add Greek yogurt to your ice cream maker. It's creamier and freezes more quickly than regular yogurt. The healthy bacteria will even survive the freezer.

FIX 3

BOOST GUT

Eat a fermented food every day.

We're conditioned to fear germs, yet there are as many bacteria as there are human cells in our bodies, and additionally, certain species of bacteria help us to digest our food, stay healthy and keep the weight off. Our digestive tract needs to house plenty of these helpful bacteria to balance out the harmful ones. Primarily, the food that we eat determines our gut microbiota profile or microbiome. Some foods feed the bad bacteria, and some help promote the health-promoting bacteria.

Fermented food

Fermented food is made by bacteria feeding on the carbohydrate in the food and releasing lactic acid. The lactic acid helps your body digest, absorb, and assimilate nutrients.

Fermented foods containing live bacteria help keep the microbes in good balance. Like wealthy tourists that spend money and help the local economy, live active bacteria found in some foods can improve your bacteria balance in your gut. Those touristy bacteria don't stick around for long, so you need to eat these foods on a daily basis to keep your microbiome healthy and boost your immune system. Gut bacteria even produce nutrients, including biotin, vitamin B12, and vitamin K.

Healthy microbiome

A healthy gut lowers intestinal inflammation, aids in digestion, and helps us avoid disease, as well as plays a role in keeping weight under control. Several studies have reported that the microbial populations in the gut are different between lean and heavy people and that when heavy people lose weight, their microorganisms become similar to those of lean people. In fact, when the gut bacteria of pairs of lean and obese twins were transplanted into mice, the mice that received the lean twins' bacteria remained lean, while the mice that received the obese twins' bacteria became fatter, even though they ate the same amount.

Brain and gut connection

The impact on brain function is becoming more apparent. Research indicates that our gut microbes affect our mental responses. Live microbes in yogurt can lead to significant improvements in depression, anger, and anxiety. Findings from UCLA research have indicated that bacteria in yogurt may actually change the way our brains respond to our environments and can even improve brain function.

Fermented food

Yogurt is full of good bacteria, so add it to your diet on a daily basis when possible. Aged cheese is often created, in part, by adding live bacteria cultures, which ferments the milk's natural sugars into anti-inflammatory lactic acid. Romano (Pecorino Romano), Gouda, some cheddars, Parmesan cheese (Parmigiano-Reggiano), and even cottage cheese are made using this method. Many of these cheeses still contain some live bacteria. Cottage cheese doesn't usually contain live strains of microbes once pasteurized but may still benefit your health from being made with bacteria.

Yogurt is by far the most commonly eaten probiotic in the United States and contains more live cultures than most other fermented foods. Make sure that your yogurt contains multiple strains of live active cultures, is low in sugar, and doesn't include thickeners or fillers such as lecithin, carboxymethylcellulose, or polysorbate-80. Kefir is a yogurt-like probiotic drink that contains even more live bacteria than yogurt. It typically contains ten different strains of helpful bacteria, leaving it 99 percent lactose-free.

Nondairy options

Both yogurt and kefir can be completely dairy-free when fermented with nondairy kinds of milk such as soy, almond, rice, or coconut milk. Most of these nondairy yogurts won't provide nearly as much protein or calcium, but they are fermented foods that can be beneficial.

Other nondairy foods, such as unpasteurized sauerkraut, olives, Korean kimchi, soy sauce, pickles (not made with vinegar but authentic fermented pickles), pickled vegetables, olives, and soybean products (tempeh, miso, and natto) are fermented. They may not contain nearly the gut-restoring probiotic strength that yogurt and kefir do, but when eaten consistently, they also boost gut health.

Because the live bacteria are sensitive to heat and time, many of these foods are void of live cultures when you buy them at the store. Look for fresh sauerkraut and olives at your deli. Kimchi is fermented and pickled cabbage mixed with other ingredients. Fermented vegetables in jars have been heat-treated, so the live cultures have been destroyed. Fresh kimchi can often be found in Asian markets and restaurants. You can also buy unpasteurized miso paste in the refrigerated section of a grocery store.

Probiotic consistency

When it comes to fermented foods, how often you eat them is more important than how many of them you eat. Eating fermented foods regularly replenishes the bacteria in our digestive systems and keeps us healthy. Be consistent, and try to eat at least one fermented food every day.

Prebiotic foods

Not only does the good bacteria in some foods (called probiotics) boost your gut-bacteria balance, but foods like fruits, vegetables, legumes, nuts, seeds, and whole grains nourish the helpful bacteria in your gut. The fiber in its whole-food form appears to be the key to nourishing the friendly bacteria in your gut. Sugar and refined, processed foods feed the bad bacteria, disrupting a healthy balance of microbiota in your gut.

Action plan

- ✓ Eat fermented foods containing live microbes regularly to replenish your gut bacteria.
- ✓ Add low-sugar varieties of Greek and Icelandic yogurt and kefir to your diet daily.
- ✓ Add your own sweetener, like pureed fruit, raw honey, or real maple syrup to plain yogurt.
- ✓ Try making your own yogurt. It's easy, inexpensive, and delicious.
- ✓ Eat lots of fruits, vegetables, and whole grains to slow down bad bacteria growth and provide food for good bacteria.
- ✓ Eat between 25–29 grams of fiber in whole foods like whole grains, legumes, fruits, and vegetables to boost your gut microbial balance and diversity.
- ✓ Include a variety of prebiotic foods in your diet on a regular basis.
- ✓ Avoid processed foods (including gluten-free products) that lead to a poorly balanced microbiome in your gut.
- ✓ Eat more plant-based protein like beans, lentils, nuts, and seeds to improve your gut microbiome.
- ✓ Track your progress on the *12 Fixes to Healthy* app.

FIX 4
PUSH PRODUCE

Think about eating fruits and vegetables as your daily "detox cleanse"!

Fruits and vegetables deliver thousands of phytonutrients that help cells stay healthy and repair DNA damage caused by pollution, poor diet patterns, alcohol, smoking, and other harmful toxins. Eating lots of fruits and vegetables gives your immune system a tremendous boost and is a critical factor in disease prevention.

Fix 4 is to eat half of your food in fruits and vegetables. That goes for both meals and snacks. Beans and lentils are vegetables too, so don't leave them out, but don't count french fries in even though they are technically a veggie!

It sounds easy, and it is, once you make your mind up that you will make it happen as often as possible. You will need to make a concerted effort to actually eat the produce. Once you do, you'll soon get in the habit of throwing produce in your purse, desk drawer, and briefcase. Before you know it, you will suddenly see vegetables on menus that you hadn't noticed before.

I'll show you ways to make vegetables taste delicious. Fruit will be more satisfying than ever before, and you'll feel better. Share your vegetable techniques with your friends and coworkers, and bring produce into work instead of donuts!

PRODUCE POTENTIAL

Packed with tens of thousands of nutrients, produce in natural form work together in powerful ways to reduce inflammation, lower your risk of disease,[1] and help you keep fat weight down.[2]

Pills can't compete with Mother Nature. Studies comparing the effectiveness of vitamins and minerals taken in supplement form versus in their natural state have shown that the supplements aren't nearly as powerful as nutrient-dense whole foods. Because all the nutrients work together synergistically in ways we don't understand, individual nutrients don't have the same impact on their own.

In 1994, research established that deep-orange fruits and vegetables high in carotenes, such as carrots, had the power to fight cancer, especially lung cancer. Researchers assumed that the beta-carotene in supplement form could reduce the risk of cancer but were surprised to discover that beta-carotene supplements not only were useless against cancer but actually increased the risk of cancer for people who smoked. Only natural carotene consumed in whole-food form proved to be a powerful ally against the disease.[3]

Produce nourishes gut microbes

The three pounds of microbes inside your digestive system are nourished by whole foods high in fiber, like fruits and vegetables. In the past decade, researchers have discovered that the microbiome is intertwined with the immune system to fight disease.[4]

An imbalance of bacteria in the gut negatively affects your weight, ability to fight off sickness, risk of disease, and even your responses to the world around you. In fact, recent technology in gut microbiome research shows that subtle imbalances in the gut microbiome impact many diseases. Your gut bacteria live off of whatever's leftover in your colon after your cells have digested all of the amino acids and nutri-

PRODUCE VARIETY

Did you know that a diet consisting of lots of different types of fruits and vegetables can lead to a healthier, more diverse gut microbiome? In fact, adding a variety of plant foods (in their whole form) to your diet can change your gut bacteria profile in just a few days.[5] So if your diet is in a rut, so is your gut. Eat a rainbow of fruits and vegetables to take in a variety of different types of produce!

ents. So eating fiber-rich foods like fruits and vegetables instead of sugary, processed foods is the best way to encourage a balanced and diverse microbiome that is so critical to your well-being.[6]

Powerful plant chemicals

The thousands of plant chemicals that protect plants from drought and disease also help to protect you from disease![7] More than twenty-five thousand phytochemicals are found in whole foods. All those thousands of plant chemicals also help to decrease chronic inflammation, insulin resistance, and the risk of chronic disease.[8]

The phytochemicals in berries and other deep blue, red, and purple foods are powerful! The anthocyanins have been shown to help keep blood sugar levels lower after eating a carbohydrate-rich meal by slowing down or inhibiting an enzyme that breaks down sugar.[9]

Brain perks

Research indicates that phytochemicals, often referred to as phytonutrients, protect brain tissue from breaking down and promote quicker and sharper cognitive skills. Specifically, vitamins C and E may aid neurons to better communication.[10] Vitamin A appears to boost the brain's ability to shift from one mental activity to another.[11]

Certain phytochemicals, including anthocyanins that are in deep-red and blue-colored produce appear to decrease inflammation in the brain, improve brain signals, and improve blood flow to the brain.[12]

A study following 13,388 aging women for twenty-five years found eating any kind of vegetable was associated with significantly less decline in thinking and memory skills. Cruciferous vegetables (broccoli, cauliflower, cabbage, kale, Brussels sprouts, and bok choy) led with the greatest brain benefits, while green leafy vegetables followed.[13]

EAT GREENS TO KEEP THINKING

One study reported that people who ate just over a cup of fresh leafy green, or just over a half cup of cooked greens had a slower rate of decline in thinking skills and memory during mental exams than people who rarely or never ate them. These older green-eating participants were eleven years younger cognitively than those who didn't indulge in green vegetables.[14]

Eat produce, lose weight

Eating produce can help you be more satisfied and want less food. The highly controlled crossover study done by the National Institutes of Health reported a strong link between eating whole foods and weight loss and more satiation.

Participants of the minimally processed, whole-food-diet group voluntarily ate an average of 508 calories less per day and lost about two pounds over a two-week period instead of gaining an average of two pounds. The whole-food diet included lots of fresh fruit and vegetables and the ultra-processed diet included canned soup, canned chicken, and other commonly used processed food items. The large calorie and weight changes occurred even though researchers made the diets equal in carbohydrates, fats, proteins, sugars, fiber, and sodium.[15]

Also, don't forget about the Harvard study I mentioned in the introduction that evaluated seemingly insignificant habits affecting the body weight of more than 120,000 people over twenty years. The people who lost weight ate lots of fruits and vegetables, and the people who gained weight didn't. Produce was identified as an important factor in body weight.[16]

I don't think I have to convince you that produce helps you lose weight and keep it off. But what I want to motivate you to do is actually choose to eat produce, not skip food altogether. The powerful nutrients in fruits and vegetables help you to be more satisfied so that you don't need to eat as much later in the day. Plus, those thousands of plant chemicals make a difference in how you feel and think!

Live longer

Eating half of your food in produce might even lengthen your life. A meta-analysis of ninety-five studies looking at fruit and vegetable intake found that 7.8 million premature deaths in the world could be avoided if people ate 800 grams of fruits and vegetables per day.[17] For reference, one medium apple alone is about 182 grams. Eating produce certainly can help improve the quality of your life, given all the health perks associated with eating them.

Daily detox cleanse

Fruits and vegetables deliver thousands of phytonutrients that are extremely powerful. The Dietary Inflammatory Index (DII) rated the nutrients in fruits and vegetables among the highest, with some of the greatest impacts in fighting and repairing damage throughout the body.[18]

Forget detox juices. Eating half your meals and snacks in fruits and vegetables in their whole form is your best daily "detox cleanse."

PRODUCE SHOPPING

We go up against a lot of advertising every day that's designed to make us crave unhealthy foods. Fortunately, we still have the power to actively decide what to buy and eat.

More people are shifting to nutritious foods. The more food in the whole form you choose to buy and eat, the more you will appreciate the delicious taste of real food. Processed foods filled with refined white flour, sugar, and salt will slowly lose their appeal. You will become more energetic. It takes some willpower not to reach for the bags of chips and cookies on the supermarket shelf during those first few months. But as time goes on, you will start to prefer whole foods to processed foods.

You have the power to change your tastes. And that starts with what you put in your shopping basket.

Produce convenience

While processed foods in the grocery stores outnumber whole foods, there are plenty of healthy foods that can make your transition to whole-food eating easier and more delicious.

Having a variety of frozen fruit sitting in your freezer all year round comes in handy. Frozen berries pureed in a blender can top healthier versions of pancakes, waffles, french toast, and plain yogurt. I like to take a handful of frozen blueberries, run them under hot water for just a few seconds, and throw them into my oatmeal or yogurt for a delicious breakfast. There's an unending list of ways you can use frozen fruit, and buying it can help you conveniently eat high-nutrient foods.

Produce frozen the same day that it is harvested retains its nutrient levels extremely well. Phytonutrients called anthocyanins, best known for their deep red and blue colors, need to be preserved in food so that we can take advantage of their powerful disease-fighting nutrients. Frozen fruit is full of these wonderful phytonutrients, along with much more.[19]

Compared to fresh vegetables, the texture of many frozen vegetables may not be as good. Grated vegetables, however, are a different story. Frozen grated vegetables, like cauliflower and broccoli, are showing up in droves in the frozen sections of grocery stores. They look and taste similar to rice and can be used to replace rice in recipes. These riced vegetables cook in minutes and taste great both alone or as a side to your meal. They may be labeled "riced," "crumbles," "pearls," or just what they are: grated vegetables that look similar to rice.

While it's easy to make these grated veggies yourself in a food processor, it's smart to have some stocked in your freezer when time is not on your side. As long as these products are just frozen grated vegetables and no other ingredients are added, they make a convenient way of getting more veggies into your diet.

Buying fresh vegetables already washed and cut means it takes fewer steps to get them onto the table. Brands like Eat Smart sell fresh vegetables that are cleaned, trimmed, and sealed in a stay-fresh package. They offer organic greens like rainbow chard, kale, and fresh salad mixes with a variety of greens you may never try on your own. A tasty premade salad can make your transition from processed foods to whole foods a breeze.

Don't spill the beans

Beans, lentils, and dried peas, referred to as pulses, are an amazing boost to your health. No other food group can claim to be from both the vegetable and the protein group.[20] Check the labels: as long as nothing else is added except some salt or spices, buy these plant-based proteins either dry, canned, or already cooked (produce section) in your stores or farmers' markets. I get the best deals in ethnic stores for these legumes in their dry form.

I enjoy snacking on roasted chickpeas. My favorite is Biena Chickpea Snacks. They use US Grade 1 chickpeas, with less oil for a nice crunch. The texture is crunchy but not too hard like some of the other brands I've tried. I'm all about eating my vegetables and protein in one snack.

Processed snacks

Processed snacks are a missed opportunity to eat produce in its whole-food form. Once kids get used to eating these types of highly processed foods (designed to be overeaten), they tend to not want whole-food snacks like a real veggie stick or an apple slice. That's often true for us adults as well. Vegetables and fruits in their whole-food form are actually difficult to overeat but, give yourself or a child a processed snack, and it's harder to stop. Over time they'll learn to listen to their taste buds instead of their stomach.

On top of the poor ingredients, processed veggie products are usually fried even though their labels are misleading. There are several studies linking the acrylamide formed when potatoes and other starchy foods are fried to increase the risk of heart disease and mortality.[21]

While some veggie chips are made with actual vegetables (and not potato starch), they may be an upgrade over regular chips but never better than the opportunity to improve your child's health and habits with vegetables and fruits in their whole form!

VEGGIE STRAW WARNING

Offering your child veggie straws, sticks, or chips may seem like a healthy substitution for potato chips or Cheez-Its, but don't be hoodwinked by the word "veggie" in the name. Veggie chips and straws are not health-promoting! They are ultra-processed foods that tend to negatively affect the microbial balance in children's guts (as well as adults) while replacing an important opportunity to feed their friendly health promoting microbes. Poor bacteria balance and diversity in the gut is being linked with most of the health issues on the rise today, from celiac disease to obesity.

Typically, veggie sticks contain potato flour or potato starch as their primary ingredient. Powdered spinach is closer to the end of the ingredient list, so an insignificant amount of processed spinach is actually added to these veggie snacks instead of vegetables. Most of these products are more processed than potato chips; potato chips at least start with a whole potato, so there is a little fiber and a few more nutrients than these veggies-stick products.

TRICKY LABELS

Labels are not always what they seem. They are great in theory. But in practice, you might be surprised by how many loopholes food manufacturers can find when it comes to making their labels say what they want them to say.

Processed foods come with different labels, but just because something is labeled "organic," "natural," or "GMO," doesn't mean that it isn't highly processed.

Minimally processed foods are nutritious foods that are simply cooked, cut, frozen, crushed, canned, fermented, or mixed. Examples of minimally processed foods are roasted vegetables, frozen fruit, nut butters without added ingredients, unsalted canned beans, plain Greek yogurt, or even homemade cookies (using raw honey, rolled oats, whole grain flour, etc.). You get the idea!

IDENTIFY HIGHLY PROCESSED FOOD

Two quick ways to identify if food is more than minimally processed or highly processed using the ingredients list:

1) Are the ingredients whole foods or are they highly processed? Does the product use whole fruit, raw honey, or whole-grain flour or processed sugar and refined flour?

2) Can you buy these ingredients to use in your own kitchen? Or are they only made in a lab, using a chemical process (like hydrogenated oil, high-fructose corn syrup, soy protein isolate, or aspartame)? If you can't make this product at home using whole-foods ingredients, then it's considered highly or overly processed.

"Organic"

Organic foods are great because, let's face it, it's nice to purchase your whole foods without an extra dose of pesticides. A Stanford University study published in the *Annals of Internal Medicine* in September 2012 reported that eating organic meat and produce reduced the consumer's exposure to antibiotic-resistant bacteria.[22]

The organic movement has generated a lot of hype and attention. "Organic" means that the food has been grown without the use of pesticides, synthetic fertilizers, sewage sludge, irradiation, genetic engineering (GMOs), growth hormones, antibiotics, or other drugs. Also, the food products have been made without food additives such as preservatives, artificial sweeteners, colorings, flavorings, and monosodium glutamate.

To be labeled "organic," products must be organically grown or made from at least 95 percent organic ingredients and 100 percent from non-GMOs. Products that have at least 70 percent organic ingredients can say, "Made with organic ingredients" on the label. These requirements are good, but they're not always the whole story.[23] *Just because something is labeled "organic" doesn't mean that it isn't processed.*

One example is white flour. White flour is wheat flour from which the bran and the nutritious germ of the grain have been removed. That means that white flour is processed, regardless of whether it was grown without using pesticides.

I encourage you to buy organic whenever you can. However, don't assume that because something is organic it must be nutritious. Processed foods are unhealthy regardless of whether their ingredients were grown with pesticides!

If you can't afford to buy organic fruits and vegetables all of the time, don't panic. The same 2012 Stanford study also pointed out that organic foods themselves may not be significantly more nutritious than conventional foods. It's still more important to trade in processed foods for fruits and vegetables. Try buying them in season from local farms for the freshest, least expensive options.

"Natural"

We see the term "natural" on food labels all the time. It sounds good, but be careful. The term "natural" actually has no legal meaning in the United States.

Many people assume that if something is labeled "natural," it is minimally processed and doesn't contain manufactured ingredients such as processed sweeteners, lab-produced "natural" flavors, and colors, additives, or preservatives.[24] But because there is no labeling standard for this word, it can be legally printed on any food label—regardless of what the package contains.

Be wary of the word "natural" on food labels, and pay close attention to the ingredient lists. If you find phrases such as "added color," "artificial flavors," or "synthetic substances" on the list, the food isn't really natural. Always try to buy foods in their whole form or products made with whole-food ingredients to truly be eating more naturally.

GENETICALLY MODIFIED FOODS (GMOS)

Genetically modified organisms (GMOs) are a controversial topic, and there are definitely two sides to the debate. On one side, the biotech industry and GMO supporters view genetically modified (GM) foods as a way to grow more food and potentially feed millions of starving people as well as keep rising food prices down.[25] On the other side, many believe that GM foods require more oversight and argue that foods containing genetically modified organisms could potentially be harmful to humans and ecosystems.[26] They push for GMOs to be outlawed, as they are in Europe, and at the very least be labeled clearly so consumers know whether foods are genetically modified.

GMO supporters claim that Americans have been eating foods containing them for the past fifteen years without credible evidence that people are harmed,[27] yet scientific clinical studies using humans are almost nonexistent. Saying that there is no evidence of harm isn't the same as saying that GM foods have been shown to be safe. Also, when Americans were recently compared with Europeans who do not allow GMOs into their food supply, there wasn't any clear evidence that GM foods are increasing the rates of cancer, obesity, type 2 diabetes, kidney disease, food allergies, celiac disease, gastrointestinal disease, or autism in our country.

Happily, GMO crops have led to a decreased use of insecticides. Herbicide usage, on the other hand, has increased as genetically modified crops are more resistant to weeds. Encouragingly, efforts to find synthetic pesticides that have high specificity for their target pests, as well as pest control that is less toxic to humans, are sure to become part of the permanent solution to this pesticide conundrum.[28]

It is important to note that the Food and Drug Administration (FDA) doesn't actually require any safety assessment of GMO crops. Any animal GMO feed studies that have found health issues, including increased inflammation, were poor quality studies.[29] Significantly, these studies are few in number compared to the number of animal studies, both experimental and observational, revealing that GM feed is actually safe and nutritionally on-par to non-GMO feed for animals.[30] There is a lack of clinical research of GMOs in humans, however. Therefore, it's a personal decision whether to eat or to avoid GM foods.

GMO shopping tips

The majority of corn, soy, canola, and sugar beets grown in the United States are genetically engineered and often used in processed foods and restaurant cooking (mostly due to the oil being used). However, if these foods are organic, you can be assured they do not contain GMOs. More and more healthier-type products using corn, soy, and canola, such as popcorn, for instance, are labeled GMO-free, and more restaurants are using only non-GMOs. A greater number of non-GM foods will be available as the demand for non-GMOs grows.

In a few years, consumers will be able to find out whether any food in the stores contains GMO ingredients because the GMO label law was signed into law on July 29, 2016. Food companies will have to show whether products contain GMO ingredients, but they won't have to print it on the package label. Instead, you may have to scan a QR code (a type of barcode) with your smartphone to get that information. Smaller food companies also have the option of providing a phone number or a website URL for more information. While the GMO information for foods isn't as accessible as many hoped, it may be enough to get more companies to avoid GMOs until studies assure their safety.

PRODUCE PREP

Many people complain about not liking vegetables. But when you cook them the right ways, those complaints all but disappear. Don't forget: beans, lentils, and dried peas are vegetables!

You can transform your veggies into delicious delicacies in many different ways: shaving, spiraling, grilling, broiling, roasting, pureeing, stir-frying, blending, and tossing. In this section, I'll take you through some of my favorite preparation techniques. Keep in mind that all these prep methods work for many fruits as well!

Legumes

Lentils, beans, and root vegetables (like potatoes and carrots) thicken foods with flavor and nutrition.

Lentils

Cooked lentils are perfect for rounding out salads during hot weather as well as hearty soups and stews during the colder months. Regardless of the season, their quick-cooking, no-soak-required attributes make them ideal for hectic weeknight meals. Make extras to add to other foods throughout the week.

Lentils are much faster cooking than beans. They are an enjoyable way to thicken stews and soups while adding loads of nutrition. Red, orange, and yellow lentils cook quickly and add flavor and color to foods but tend to break down when heated—which makes them perfect thickeners. Green or brown lentils, meanwhile, tend to hold their shape better with heat.

The most reliable way to cook tender lentils is to bring them to a rapid simmer then reduce the heat to low for the rest of the cooking. They should cook with some bubbles in the water and gentle movement of the lentils. Try tossing warm lentils with good olive oil and vinegar with a cooked egg to complete the meal.

Beans

Beans can often be used to replace flour and meat in recipes. One cup of beans replaces one cup of flour in some recipes. Yes, really! If you replace beans with flour, the ingredients would need to be run through a food processor, however. Check out my Peanut Butter Chocolate Swirl Cookies to try it out at www.FoodsWithJudes.com.

TAKE THE GAS OUT OF BEANS

Make your beans easier to digest without the gas while absorbing more nutrients and improving the texture of the beans. Kombu seaweed contains enzymes that help break down and neutralize difficult-to-digest oligosaccharides in beans. Add a three-by-five-inch strip of dried kombu seaweed to the water of a slow cooker or pot before you cook the beans. Once the beans are tender, pour off the kombu floating on the top of the water into the sink and down the drain. Throw the rest of the water full of difficult-to-digest starches down the drain too! Now you can enjoy your beans and take advantage of all their health benefits!

Keep dried kombu in your pantry. It keeps for several years in a cool, dry place. You can find it in the Asian food section at many mainstream grocery stores as well as at Whole Foods, Asian markets, and many gourmet markets. Don't confuse it with sushi seaweed.

Cook beans in a slow cooker

Low-sodium canned beans work in a pinch. Make sure you rinse them at least twice before you use them. Alternatively, you can use this easy method to cook dried beans. Add the dried beans to a slow cooker (one pound or less dried beans for a three-and-a-half-quart cooker or smaller; two pounds or more for a five-quart or larger cooker).

Pour water over the beans until they are covered by at least two inches. Add about a three-by-five-inch strip of dried kombu seaweed to make the beans easier to digest with a creamier texture. Cook on low overnight (about seven hours) for most beans, but on high overnight for chickpeas. (In fact, I cook the chickpeas for at least twelve hours on high so that my hummus is extra creamy.)

Check in the morning to see if they need more cooking time. When the beans are cooked through, discard the seaweed that is floating on the top by pouring out some of the water and the seaweed. Then using a colander, strain the rest of the water from the beans. Cooked beans will keep well in the fridge for a week or in the freezer for up to three months, so you can have them on hand to add to whole grains such as brown rice or quinoa.

LECTIN LIES

While lectins and other "antinutrients" in beans (and whole-grain) have been demonized as the enemy in some popular diets recently, cooking these foods actually removes most of the lectins.[31] Uncooked beans do contain large amounts of lectins and if not cooked enough can be extremely difficult to digest.

There are specific cooking methods to eliminate more lectins so you can absorb more nutrients with fewer digestion problems. Then you can eat more of these pulses and take advantage of their amazing disease-fighting benefits for your gut-microbiome balance.

Pressure cooking beans for seven and a half minutes, for instance, deactivates almost all the lectins. Cooking kombu seaweed with your beans in a slow cooker absorbs lectins so that the "antinutrients" go down the drain.

Edamame

Another particularly delicious and nutritious bean is edamame. Edamame is simply unripened green soybeans. They have 30 to 45 percent more protein than other protein-rich members of the bean family, and unlike beans and lentils, edamame has all the essential amino acids.

Purchase edamame frozen or already cooked, with or without the pods, and add it to your whole grains, salads, and stir-fries. It also makes a wonderful high-protein snack: just cover and heat the edamame in the microwave. The beans slip right out of the pods and into your mouth; even kids have fun eating them. Best of all, there's no need to worry about negative health effects because it turns out that soy is exceptionally healthy, especially when it's unprocessed.

Processed soy

The big problem with soy is that it is often overly processed, losing many of its healthy traits. The powerful phytonutrient isoflavones is found in soy foods that aren't highly processed, like edamame. Highly processed soy foods such as soy ice cream, power bars, and hot dogs have much lower amounts of isoflavones.

Many of these products are made with soy isolates. These isolates are formed by isolating the protein from the soybean and taking out the hundreds of thousands of other nutrients that are meant to work together for your health.

Check your label to see what type of soy is included. Organic products made with soy are from non-GMO soybeans but that doesn't mean they aren't highly processed. Try eating soy in a whole-food form.

Tempeh, natto, and miso are all whole food forms of soybeans that have been processed or fermented by nature using good bacteria. That's the kind of processing you actually want to improve your gut health.

> **THE BIG SOY QUESTION**
>
> Rodents react very differently to soy than do humans. Consequently, the previous scientific research based on rodent studies, which found that soy causes breast cancer especially in those who have already beaten breast cancer, has been replaced by human studies, which found that the opposite is true in humans.[32]
>
> Edamame, a whole food without any processing, is also an excellent choice to take advantage of soy's quality and ample plant-based protein. Fermented and less processed soy is loaded with health-boosting nutrients from vitamins, minerals, fiber, and phytonutrients. This is an excellent way to transition to a more plant-based diet.

Shave your vegetables

A smoother look and texture makes vegetables more appetizing. You can shave your vegetables lengthwise with a simple vegetable peeler to create flat, long-ribboned veggies that are enticing. Plenty of vegetables are great candidates for shaving. Try it with cucumbers, carrots, parsnips, zucchini, and other types of summer squash. You can toss shaved vegetables into a salad or sauté them into silky wide noodles.

Spiral your vegetables

It takes just a few minutes to transform your vegetables into thin, uniformly curly spirals or long vegetable strands with a vegetable and fruit spiral slicer. When you do, you can create new looks and textures with almost any vegetable (or fruit) quickly with very little cleanup! Zucchini is the most popular vegetable to put through a spiralizer to make substitute pasta or noodles (zoodles), but don't stop there. Spiraled sweet potatoes make deliciously attractive noodles too. Add them to Asian soup, for example. Even spiraled rutabagas make great-tasting noodles. I love spiraled carrots and red peppers in my stir-fries and spiraled cucumbers, apples, and roasted beets in my salads. Roasted spiraled sweet potatoes or apples make the best chips. Get creative!

Grill your vegetables

A fantastic way to cook vegetables is to grill them. Grilling is one of my favorite ways to prepare veggies because the slight charring brings out some delicious caramelized flavors. Marinate vegetables in high-heat oil mixed with herbs, garlic, or lemon juice for about an hour to help caramelize them, or coat them very lightly with olive oil using an oil mister. A combination of herbs and spices such as rosemary or thyme will make your vegetables stand out. Slice soft vegetables such as zucchini a quarter-inch thick (with harder vegetables slice slightly thinner to prevent burning). You really can't go wrong with this one!

Broil your vegetables

If heating up the grill adds too much to your day, just broil your soft vegetables for a similar caramelized experience in less time. Quickly cube vegetables and toss them in a bowl with vinaigrette or olive oil, salt, and pepper. Place them on a foil-covered pan about four inches from the heating element in your oven set to medium-high. Rotate, shake, or flip the vegetables to promote even browning while you stand guard to assess their cooking state. Make sure you keep an eye on them. Everyone has tried to leave for just a second and then returned to burnt food!

Roast your vegetables

Roasting vegetables is incredibly simple. All you need is high-heat oil (avocado oil or regular olive oil), salt, and a hot oven. The vegetables take almost no preparation, and they're delicious. Roasting brings out the natural sweetness in your vegetables. They become crisp and caramelized on the outside but soft and juicy on the inside. Sweet potatoes, purple potatoes, carrots, red bell peppers, beets, onions, zucchini, butternut squash, mushrooms, cauliflower, broccoli, asparagus, Brussels sprouts—almost any vegetable you can think of, you can roast!

Use a mister to spray the oil over your vegetables or toss them in a bowl with a little high heat oil (not extra-virgin olive oil), salt and pepper, and maybe balsamic vinegar. Roast your vegetables in a hot oven, between 375°F to 500°F, until they're browned and tender. Dense, low-moisture vegetables such as root vegetables and winter squashes need lower heat and more time in the oven, while vegetables with more moisture, such as eggplant, zucchini, and broccoli, need the higher end of this temperature range. The thicker the vegetable, the longer it will take to cook.

Shake or turn the vegetables over halfway through to caramelize both sides. If they are tender but not charred, brown them with the broiler for a minute or two. Then deliver them straight to the table. For an easy meal, top your roasted vegetables with a couple of fried eggs or a dollop of plain Greek yogurt seasoned with fresh herbs.

Roasted broccoli for a win

If you haven't tried roasted broccoli, you may want to try it first. It turns vegetable haters into vegetable lovers!

Preheat your oven to at least 450°F, but up to 500°F works great. Break up and cut broccoli to a similar size, more or less. Mix the cut broccoli in a bowl with a couple of tablespoons of regular olive oil (not extra virgin) or avocado oil and salt and pepper. Cover a large sheet pan with parchment paper and spread the broccoli out on it so the broccoli isn't sitting too close to each other (otherwise, they steam each other instead of roast). Cook broccoli for about fifteen minutes in the oven. It will be browned on the edges and delicious. Enjoy!

Puree or cook down your vegetables

Pureeing or cooking down your vegetables thickens and adds flavor to a dish. Root vegetables such as sweet potatoes and carrots, and winter squash, cauliflower, beans, and lentils, or nearly any other vegetable can be added to your sauces, soups, and stews to cook down and thicken your recipes.

You can add vegetable stock or water with onions and garlic to help the process get started if needed. You can speed up the process by pureeing the vegetables (raw or cooked) before adding them to your recipe. Skip thickening with flour, cornstarch, or arrowroot if possible and try adding vegetables instead to add nutrition, flavor, and body at the same time! Use frozen or fresh fruit to replace syrup.

Stir-fry your vegetables

Stir-frying is a tasty way to eat more vegetables, but I am often asked how to make the stir-fry sauce. The trick is knowing a few key basic sauce ingredients and then tweaking the basic sauce recipe to make your favorite type of stir-fry. My formula is provided below. Combine your sauce with a variety of vegetables and protein choices you have on hand.

KITCHEN FORMULA PREP

Have you ever watched your grandmother or mother whip up an amazing dish without a recipe? Well, that's because cooks see trends in their recipes, and over time those trends become their formulas. I've provided a few formulas that make healthy recipe prep a breeze!

Simple stir-fry formula

1. 1 tablespoon high-heat neutral oil or peanut oil
2. ¼ cup combination of garlic, ginger, and onions (also chilies for more spice)
3. 4 to 8 oz. protein

    If you use chicken in the stir-fry, marinate the chicken in egg white (1), sesame oil (about 1 teaspoon), cornstarch or arrowroot (about 1 tablespoon), and rice wine (enough to cover the chicken). Be sure to drain the marinade off before you cook it to help it brown.

4. 1½ lbs. vegetables

    Ideally, use one prominent vegetable and one or two secondary vegetables.

5. Basic stir-fry sauce

Once your sauce is put together and the vegetables are cut, you are ready to go. Using a wok or large skillet over high heat, add any combination of garlic, ginger, and onions in hot oil. Cook for about a minute while stirring. Add a protein (tofu, bean curd, chicken, shrimp, etc.) and stir-fry until browned (or pink for shrimp) and just cooked through. Remove protein to a plate and cover with foil. Add the thickest vegetables to the pan first and stir-fry for a couple of minutes (you may need another teaspoon of oil if it is sticking). Then add thinner, faster-cooking vegetables. Stir-fry for about a minute. Add the sauce of choice and stir until tender but still crisp. Add the protein back in, heat through, and serve. Finish with a few toasted sesame seeds (optional) and serve with brown rice.

Basic stir-fry sauce

1. 2 tablespoons reduced-sodium soy sauce (or tamari sauce)
2. 2 tablespoons rice vinegar
3. ½ cup stock (no-salt or reduced-salt)
4. 1 tablespoon cornstarch (or arrowroot starch)
5. ¼ teaspoon (or more) red pepper flakes
6. Flavor

    Sesame Sauce:

    - 1 teaspoon sesame seed oil

Sweet & Sour Sauce:

- 2 teaspoons tomato paste (double concentrated)
- 3 to 4 teaspoons raw honey (or sugar)

Asian Lemon Sauce:

- reserve lemon zest of one lemon
- replace rice vinegar with ¼ cup lemon juice
- 1 tablespoon raw honey (or sugar)
- 1 teaspoon sesame seed oil

Shake or whisk to make basic stir-fry sauce. Choose a stir-fry flavor and make this basic stir-fry sauce come alive by following the flavor instructions.

Simple salad formula

A lot of people like green salads but don't prepare them often enough. That's unfortunate because salads can deliver a big boost to your health! Making salads and salad dressings can actually be quick and simple. Follow this simple salad formula to add more salads to your life:

1. Greens

I recommend crisp lettuce such as Romaine for at least some of the greens, and you can also add fresh basil, mint, or cilantro for extra flavor.

2. 1 to 3 vegetables

Thinly sliced red onions are a great standard vegetable to add, along with others to suit your taste—even cooked ones! Roasted or grilled vegetables add so much to a salad. Marinated vegetables, like artichokes, work too.

3. 1 to 2 fruits

Avocados are a great idea, and you can also add sweeter options such as pomegranates, pears, apples. oranges, or berries.

4. 1 to 2 protein food

Cheese, shelled edamame, chickpeas, lentils, and salmon are all great options, but any protein works. Try leftover protein from the previous night.

5. 0 to 1 whole grain

Larger, chewier grains such as farro and wheat berries are a great combination with salad. Red quinoa and black rice add color, but any whole grain works.

While whole grains are optional, they do add another dimension to a salad that keeps you full longer.

6. 1 crunch food

For the "crunch food," toss in a few toasted nuts or seeds; avoid croutons! I usually buy my nuts already toasted from Trader Joe's, but it's easy to toast nuts up in a skillet. Sunflower seeds, hemp seeds, sesame seeds, and pomegranate seeds are good seeds to add.

7. Salad dressing

Finally, add a small amount of salad dressing to the dish just before you serve it. Go easy. You can add more if you need it. Strong cheese, some protein choices, and marinated vegetables add lots of flavors, so less dressing is needed.

MASON JAR SALADS TO GO

Looking for an easy way to eat more salad on the go? Layer your salad into a wide-mouth mason jar or another similar container to keep the ingredients separate until you eat them. Premake your salads so that you're prepared to grab them and go when it's crunch time. Put your salad dressing and firmer vegetables at the bottom of the jar, giving the veggies time to marinate while your lighter, more fragile ingredients sit on top. Then just keep the jar upright until it's time to eat. Shake or stir it up, and enjoy!

Quick and easy salad dressing

I have a quick and easy formula for salad dressing too.

1. 1 part extra-virgin olive oil

Infused extra-virgin olive oil adds even more flavor.

2. 1 part white or red balsamic vinegar (or a combination of both)

Infused balsamic vinegar adds even more flavor. Note: If you use another type of vinegar besides a balsamic, the ratio to oil and vinegar changes to 1 part oil to ⅓ part vinegar.

3. **About ½ to 1 teaspoon (or more if it needs thickness) emulsifier**

    Add an emulsifier such as mustard, tahini, plain yogurt, ground flax, ground chia seed, or mashed avocado to thicken. Add more emulsifier, if you want the dressing thicker.

4. **Salt and pepper to taste**

Shake to mix.

Optionally, you can top it off with 1 to 2 teaspoons of flavorings—dried herbs or seasonings, fresh herbs, crushed garlic, or shallots.

Smoothie bowl formula

Smoothie bowls are a great way to boost your intake of fruits and vegetables. When a sweet treat is called for, they can definitely fit the bill. A good smoothie bowl is healthy, filling, and refreshing all at the same time.

This formula makes smoothies on the thicker side for the perfect smoothie bowl. Just pour your smoothie into a bowl and top with low-sugar granola, fruit, nuts, seeds, and fun surprises like unsweetened coconut. Chewing your food, rather than drinking it, may add to your level of satisfaction. Having nutritious, chewable food on your smoothie bowl allows your body to do more work breaking the food down. High-fiber foods help slow down the digestion of the smoothie bowl and burn calories more efficiently.

I have a formula for spectacular smoothies. It goes like this:

1. **½ cup liquid**

    Your liquid can be any liquid like milk, almond milk, pureed fruit, or 100 percent fruit juice.

2. **1½ cups frozen fruit**

    Any frozen fruit works. Using frozen fruit instead of ice gives the smoothie more flavor. Frozen fruit is always available and keeps nutrients longer than in fresh fruit.

3. **1 cup greens**

    Smoothies are a great way to sneak veggies such as kale (with or without the stem), baby spinach, and Swiss chard into your diet.

4. **½ cup protein**

    This can be yogurt, cottage cheese, silken tofu, nuts, peanuts, or nut butter. You can add a couple of tablespoons peanut butter powder or a protein powder, but I'm trying to encourage you to eat your food in whole-food form so all the nutrients in that food can work together to improve your health.

5. **1 fresh fruit**

    Smoothies are also a great way to use up overripe fruit. If I see an apple or another fruit looking like its time has passed, in it goes.

6. **Toppings as desired**

    Toppings: I'm talking about things like low-sugar granola, muesli, chocolate nibs, unsweetened coconut, pumpkin seeds, pomegranate seeds, berries, and other diced fresh fruit, hemp seeds, chia seeds, ground flaxseeds, sunflower seeds, nuts or nut butter, and even yogurt.

Feel free to add ¼ to ½ cup of rolled oats to this formula as well. Oats add nutrition and make the smoothie or smoothie bowl more filling.

Once you've gathered your ingredients, put them into a high-powered blender in the order listed in the formula to get the best smoothie results. Add more fresh fruit or liquid if you like your smoothies thinner. Add small amounts of a healthy sweetener (real maple syrup, pitted Medjool dates, etc.) if you want it sweeter.

If you're going to make smoothies or smoothie bowls, it's best to invest in a high-powered blender. Regular blenders burn out fairly quickly when you use them a lot, and they function much more slowly when you throw in kale, spinach, and whole frozen fruits. Even though it costs more upfront, you'll save more in the long run when you start blending your overripe fruits and veggies instead of throwing them out.

FIX 4

PUSH PRODUCE

Make fruits and vegetables 50 percent of your meals and snacks.

This is one of the most important fixes you can make for your health, and it's easy to do if you make a point of including produce in every meal and snack that you eat. Eating fruits and vegetables is your daily "detox cleanse."

Daily detox cleanse

Fruits and vegetables deliver thousands of phytonutrients that are extremely powerful. The Dietary Inflammatory Index (DII) rated the nutrients in fruits and vegetables among the highest, with some of the greatest impacts in fighting and repairing damage throughout the body.

Phytonutrients in fruits and vegetables serve as antioxidants to help cells stay healthy and repair DNA damage caused by smoking and other factors such as pollution, poor diet patterns, alcohol, and other harmful toxins. Eating plenty of fruits and vegetables gives a tremendous boost to your immune system and has been identified as a critical factor in disease prevention.

Gut boost

Fruits and vegetables feed the good bacteria in your gut! Whole foods high in fiber like fruits and vegetables nourish the microbes inside your digestive system. An imbalance of bacteria in your gut negatively affects your weight, ability to fight off sickness, risk of disease, and even your responses to the world around you. Recent technology shows that subtle imbalances in the gut microbiome impact many diseases. Eat fiber-rich foods like fruits and vegetables instead of sugary, processed foods (like veggie straws, Cheez-Its, and french fries) to encourage a balanced and diverse microbiome.

Brain boost

Our brains benefit from eating lots of fruits and vegetables too. Research has indicated that phytonutrients protect brain tissue from breaking down to promote quicker and sharper cognitive skills. Specifically, vitamins C and E may aid neurons in the brain to communicate better with each other. Vitamin A appears to boost the brain's ability to switch from one mental activity to another.

Health bang

Eating more fruits and vegetables enhances weight loss, lowers blood pressure, decreases heart disease and cancer, offers illness protection, and promotes clearer thinking. There's no question: fruits and vegetables are the biggest bang for your health!

Eat produce

Make fruits and vegetables half of the food you eat. That goes for both meals and snacks. Remember that we're not counting french fries as vegetables. Potatoes are a vegetable, but purple potatoes and sweet potatoes are even better choices. Beans and lentils are also on the list, along with starchy vegetables.

Try to include a variety of fruits and vegetables in your diet, and find new ways of preparing the ones you're used to eating. The less boring they are, the more you'll eat them!

Action plan

- ✓ Eat a variety of fruits and vegetables for half of all your meals and snacks.
- ✓ Bring fruits or vegetables with you everywhere you go.
- ✓ Keep cut-up vegetables and fruit at eye level in the refrigerator.
- ✓ Try a new fruit or vegetable at least once a month.
- ✓ Make a batch of beans in the slow cooker with kombu seaweed to make beans easier to digest.
- ✓ Try roasting your vegetables.
- ✓ Use spiralized vegetables instead of pasta more often.
- ✓ Prepare extra vegetables, including lentils and beans, to mix with whole grains all week.
- ✓ Place cut-up vegetables and fruit within arm's length.
- ✓ Prior to a meal, place cut-up vegetables out for those who are hungry while making other food items off-limits.

- ✓ Serve and eat vegetables first at your meals.
- ✓ Adopt a policy for you and your family of eating only fruits and vegetables while participating in activities such as surfing the net, checking email, reading, or watching TV.
- ✓ Track your progress on the *12 Fixes to Healthy* app.

FIX 5
EAT EARLY

Up to this point, we have been discussing WHAT to eat. This fix is about WHEN to eat, which matters more than you may realize.

The cells in your body need time to repair and restore during the night when your digestive system is quiet. A growing body of evidence suggests that an overnight fast, aligned with darkness and your body's nighttime response, may be a critical piece of the weight-loss puzzle and healthier living.

Fix 5 is to eat during the day so that you fuel during the daytime and stop eating early in the evening.

EAT ACCORDING TO YOUR CIRCADIAN CLOCK

You essentially have a daily clock in your brain along with clocks in the various organs of your body, which direct daily activity.[1] Together these clocks make up your circadian rhythm, which is controlled by light and darkness.[2] Light starts metabolic, physiological, and behavioral programs that function on a 24-hour cycle.[3] Just as your brain needs sleep each night to rest, repair and rejuvenate, all your other organs need downtime to reset as well.[4]

Morning eating

Eating food in the morning is in sync with these programs, which begin when you wake up and continue during the day.[5] In the morning, the insulin in your body is most effective at doing its job of moving glucose (blood sugar) out of the blood and into the cells, where it can be used for energy.[6] Insulin becomes less effective at shuttling blood sugar as the day progresses and is the least effective at night.[7]

Nighttime eating

Eating food at night conflicts with nighttime pathways that encourage sleep rather than eating. Some of your body's hormones are active at night, such as melatonin and growth hormones, and interfere with insulin's activity.[8]

If you eat ice cream at night, for instance, melatonin actually blocks some of the action of your insulin so that sugar from the ice cream stays in your blood for a longer period of time. Higher blood sugar levels mean more insulin is produced to do the job, which tends to put you in fat-producing mode.[9]

> **SWEET DREAMS**
>
> Late-night food can cause indigestion and interfere with sleep. Your body shifts from eating mode into sleep mode during the evening. Keep your evening meal light and skip those nighttime snacks to improve your sleep. Researchers found that people—especially women—who eat during the evening had lower quality sleep.
>
> The same goes for alcohol. While a drink of alcohol might help you relax before bed and initially fall asleep, drinking even two alcoholic drinks causes you to wake up more frequently, robbing you of REM and deep sleep. Alcohol-related sleep issues are even worse for women.

Weight loss: timing versus calories

There is research to demonstrate that calories eaten later in the day promote weight gain. Harvard professor Dr. Scheer reported in the International Journal of Obesity that dieters who ate their main meal before 3:00 p.m. lost significantly more weight than those who ate later in the day. This study showed that if you eat at least two-thirds of your calories before dinner, you'll consume fewer calories over the whole day.[10]

Time beats calories

The timing of eating appears to be more important than overall total calories. In another study published in the same journal, two groups ate the same number of calories. Overweight women who ate a big breakfast (seven hundred calories) and a light dinner (two hundred calories) lost more than twice as much weight as did women who ate two-hundred-calorie breakfasts and seven-hundred-calorie dinners.[11]

Not only do night owls tend to consume more calories, but they also may fail to burn as many calories as those who go to sleep earlier.

BURN MORE FAT WHILE YOU SLEEP

Your body readily uses carbohydrates for fuel. The sooner in the evening that they use up those carbs for energy, the sooner your body starts burning body fat while you're sleeping.

There is evidence that eating later in the evening increases body-fat storage during sleep, regardless of exercise or food choices.[12] A recent study backs up this growing association between late-night eating and higher body fat in humans. Three types of technology were used in this unique study to record participants' sleep, physical activity, and eating patterns. Eating later in the evening was associated with greater body-fat stores.[13]

When you time most of your eating to occur earlier in the day, you stand a better chance of losing fat weight instead of gaining it.

> **GUT BOOST**
>
> The three pounds of microbes inside your digestive system have a circadian rhythm too, and they need a break at night from their labors of digesting food. When these microbes can reset and restore the cells in your intestinal wall during the evening and overnight, the diversity and balance of your gut microbiome improves. That often translates into improvements in weight and a stronger immune system, ready to fight off illness.

An argument for breakfast

Break your overnight fast! Skipping breakfast may seem like a good way to work in a longer fast, but morning is when your body metabolizes carbohydrates most effectively. Instead, stop consuming food early in the evening when your body is more insulin resistant.[14]

Intermittent fasting

Intermittent fasting has been a hot topic lately. Early time-restricted feeding (eTRF), or prolonged nightly fasting, is a form of intermittent fasting that involves eating early in the day and ending your eating early in the evening to align with your circadian rhythm. Most of the evidence in humans indicates that eating the majority of your food earlier in the day is better for your health.[15]

Intermittent fasting carried out later in the day is in opposition to your circadian rhythm. Your rhythm is not only set by your exposure to bright light throughout the day but also by the food you ingest.[16] Eating late in the evening sends conflicting signals to your body. Fasting through breakfast and eating late into the day and evening may be problematic, especially in terms of insulin sensitivity[17] and inflammation.[18]

Protein-rich breakfast

A higher-protein breakfast appears to decrease hunger levels throughout the day.[19] Breakfast that includes enough protein that you eat—not drink—tends to help control nighttime eating by suppressing evening food cravings.[20]

Dr. Heather Leidy, lead researcher and assistant professor in the Department of Nutrition and Exercise Physiology at the University of Missouri reports a breakfast containing at least 30 grams of protein is the best way to achieve fullness and curb food cravings. Brain scans done on participants in Dr. Leidy's many studies have shown that eating more protein in the morning decreases activity in the reward-driven, pleasure-seeking part of the brain associated with food cravings later in the day and even at night. Even better, Dr. Leidy and her team have found that high-protein breakfasts increase activity in the executive decision-making part of the brain.[21]

So eating a more substantial high-protein breakfast enables you to eat less at night, in line with your circadian clock.[22]

LOSE INCHES

Researchers are finding that a longer lapse between dinner and breakfast helps to decrease body fat stores. According to a frequently cited study, mice that ate an early dinner and then fasted for 16 hours were slimmer than those who ate the same number of calories per day but snacked around the clock. Although more research is needed, this appears to also be the case in humans.[23] Just another reason for you to say goodbye to late-night snacking!

Skipping breakfast

In studies, overweight adults lost more weight when they ate a large breakfast, a modest lunch, and a small dinner.[24] Skipping breakfast can also increase your risk of disease.[25] Chronic diseases such as heart disease and type 2 diabetes are of particular concern.[26]

If you want to extend your fast, begin fasting earlier in the evening rather than skipping or delaying breakfast.

BENEFITS BEYOND WEIGHT LOSS

Overnight fasting followed by breakfast is also associated with improved blood sugar (glucose) control.[27] But what about the risk of cancer? One particular study that followed more than three thousand breast cancer survivors who underwent a nightly fast of more than thirteen hours showed lower cancer recurrence, better glycemic control, and even improved sleep.

Blood pressure may also be improved by following early time-restricted feeding (eTRF). Both blood pressure and oxidative stress were found to decrease in a tightly controlled, crossover study with prediabetic men. This research demonstrated that intermittent fasting's beneficial effects are not due solely to weight loss and that eTRF may improve some aspects of cardiometabolic health.[28]

Ending calorie consumption after 6 to 7 p.m. and even earlier when possible, is a pattern you may want to consider adopting, regardless of your need for weight loss. Your health and sleep may also improve.

IMPROVE HEALTH

Did you know that if you eat late at night, you tend to eat more? That's not all. Studies reveal that late-night munching increases fat in your blood, called triglycerides. When you eat, your body converts any calories it doesn't need for energy into triglycerides; and high levels of triglycerides may increase your risk of stroke and heart attack. A group of researchers at Columbia University Medical Center in New York interviewed nearly thirteen thousand people participating in a national study investigating Latinos and Hispanics' health.

The more participants ate after 6 p.m., the higher their blood sugar levels. Those who ate about a third or more of their calories during the evening had a 23 percent higher risk of high blood pressure.[29]

BREAKFAST IS THE NEW DINNER

Work with your body, not against it. Eat food during the day to match your circadian rhythm. Start your morning off with a health-promoting, high protein breakfast to be more satisfied and eat less in the evening. Don't be tempted to skip or delay breakfast just to lengthen your fast. Daytime is when you need fuel to be the most productive and active. Nighttime tends to be more sedentary, with less need for food for energy.

Dinner

While eating earlier in the day appears to be better for your metabolic health, that does not necessarily mean that you should skip dinner. Eat the majority of your food during the day combined with a relatively light dinner. Overnight fasting while eating whole foods instead of processed foods can help you eat less in the evening as well as help you lose weight more effectively.[30]

So yes, it does matter when you eat and drink! You metabolize the carbohydrates in your food far better in the morning soon after you wake up compared to at night when your sleep hormones kick into gear. Eat breakfast like a king, lunch like a prince, and dinner like a pauper!

EAT, DON'T STARVE!

Warning! This fix is not about skipping meals or cutting calories; it's about eating plenty of whole foods during light time hours to fuel your body so you can think and move. Avoiding food should only happen during a twelve- to fourteen-hour period of time overnight when you don't need fuel and while your body resets, repairs, and burns fat stores for energy.

It's also about eating enough of the right foods during the day to keep you satisfied, help you lose weight, and improve your health. It's so easy to think that if you eat less you will lose more weight. You might lose more weight initially, but the weight you lose at first will primarily be water and muscle weight instead of fat weight. The minute you eat normally, water and muscle weight will return. It's really fat weight that you want to lose anyway.

To lose fat weight, you actually need plenty of healthy protein foods, half your food in produce, a few whole grains (more if you're working out a lot), some healthy fats, and plenty of water during the day. You will not be hungry, that's for sure, but you will boost your metabolism and lose fat weight. Plus, eating whole foods will keep you satisfied enough to easily skip nighttime food and will give you the energy you need to move more and feel amazing!

KETO DIET

If you've ever considered going on a keto diet, think again. You can lose just as much weight using this fix, by eating early and ending your eating early enough that you fast for twelve to sixteen hours overnight. As a wellness dietitian, I want to point out a few of the hidden costs of the ketogenic diet in contrast to some of the benefits of an overnight intermittent fasting regime. Just hear me out on this one.

Keto diet overview: low carb, high fat

The ketogenic diet, also called a ketosis diet or keto for short, is a way of eating that mimics the effects of fasting. During fasting, body fat is metabolized, which makes ketones that the body uses for fuel rather than glucose from food.

In an attempt to force the body to use ketones for energy on a ketosis eating plan without fasting, dietary fat replaces carbs in the diet. Thus, the keto diet is very high in dietary fat (typically 70 to 80 percent of your total calories), while very low in carbs (less than 5 percent). This low-carb intake eliminates refined grains and sweets from the diet, along with many vegetables, fruits, beans, lentils, and whole grains.

Without carbohydrates to use for energy, your body is forced to use fat to produce ketones as the primary energy source. On a ketosis diet, the primary source of fat used to produce ketones is from the fat you eat, not body fat as occurs in fasting.

Many assume that protein consumption is unlimited when following a ketogenic diet. But since the body converts extra protein to glucose to use for energy, only moderate amounts of protein (about 15 to 20 percent of total calories) can be eaten to keep your body burning ketones for energy.[31]

Valuable nutrients lacking in a keto diet

Eliminating whole food groups from your diet is a dietitian's nightmare! Why? Because the body needs these critical nutrients to function properly. Avoiding a variety of fruits, vegetables, beans, lentils, whole grains, and specific dairy products (like yogurt without added sugar) that contain "good carbs" has significant health impacts over the long term such as suppressed immunity, poor gut and bone health, risk of dehydration, and many more.[32]

The scientific research is full of studies pointing to the importance of eating half of your food in fruits and vegetables to decrease the risk of disease, boost your immune system, and even help preserve your thinking and memory skills. Yes, that's right; 50 percent of the food you eat should be a variety of produce.[33] Fruits and vegetables are full of not only vitamins and minerals but also fiber and thousands of phytochemicals that work together synergistically—in ways we don't yet understand—to improve your health and weight.

With limited fruits and vegetables, whole grains, and legumes, people miss out on fiber, specific vitamins, minerals, and phytochemicals that only come in these foods and that are essential for gut health.[34] Not only do these foods feed the good bacteria in your gut to improve your microbiome but constipation commonly results without these valuable carbohydrate foods in your diet.

Important electrolytes, such as sodium, magnesium, and potassium, are found in fruits, beans, starchy vegetables, and yogurt, which are restricted on a ketosis eating plan. In addition, ketosis causes you to urinate more and lose valuable electrolytes. It's no wonder the body is oftentimes low in electrolytes and subject to dehydration when following a ketogenic diet. Besides dehydration and lightheadedness, keto dieters are prone to kidney injury and kidney stones.[35] Electrolytes are critical for the heart to beat correctly, so ketosis dieters may be at risk for cardiac arrhythmia too.[36]

Since ketones produced during the breakdown of fat are acidic, the blood becomes acidic, and the body compensates by pulling minerals from bones. Over time, calcium loss can lead to increased fractures, low bone density, and osteoporosis.[37]

Keto diet and food shaming

While following a ketogenic eating pattern and avoiding so many health-promoting foods, it's easy to begin to think of eliminated foods as "fattening" and "bad" for you. Feeling guilty about eating a highly-nutritious apple or a sweet potato is problematic for your health. Your body needs those health-promoting foods to help you think and function. But even when people have stopped following the diet, there is often a continued sense of shame when eating nutritious carbohydrate-containing foods such as fruit, many vegetables, whole grains, and yogurt. At the very least, there is a lingering mindset that these highly nutritious foods are to be avoided.

The keto plan focuses on how much and what can be eaten rather than eating to satisfy hunger cues and eating quality foods in their whole-food form. There is plenty of research that indicates that the key to successful and sustained weight loss is more about eating whole foods instead of processed, refined foods.[38] Yet, one of the main problems of following a very low-carb diet is that people tend to eat a lot of processed protein and poor-quality fats, with very few fruits and vegetables and whole grains. Just as healthy carbohydrate food may be viewed by a ketosis dieter as off-limits and "bad," these less healthy processed foods may potentially be seen as health-promoting and "good" for weight loss.

Keto diet obstacles

While your body adjusts to burning fat rather than carbs, the transition is often uncomfortable. Referred to as a "keto flu," many people experience symptoms such as headaches, fatigue, nausea, irritability, hunger, and brain "fog" lasting about a week.[39] These unpleasant conditions do gradually decrease as your body gets used to converting ketones into energy, but during that time, it's harder to be productive, function at work, and exercise.[40]

Ketosis adherence and sustainability

The rules on the ketosis plan are so rigid (for your body to be able to convert to burning ketones for energy) that it's easy to throw in the towel with one slip up. It takes a few days for your body to achieve ketosis. But if you "cheat," then your body is no longer in ketosis, and it takes another few days for your body to reestablish ketosis. This diet is difficult to adhere to, so it's not surprising that the *US News & World Report*'s review of 2018 diets rated the ketogenic diet in the last place as a sustainable means to weight loss.[41]

Because of the keto diet's challenging nature, low adherence rates result. Quitting the diet can lead to binge eating to satisfy food deprivation, which can result in weight gain. An all-or-nothing dieting mentality often develops when dieters find themselves unable to adhere to a diet. When it comes to sustainability, the keto plan is a lose-lose!

KETO DIET MICROBIOME WARNING

Following a keto diet or any type of low-carb diet may help you lose weight in the short term, but your gut balance and diversity of microbes may be left primed for weight gain. Restricting plant-based carbohydrates like fruits, vegetables, whole grains, beans, and lentils can have a major negative impact on the gut microbiota.[42]

Fiber from whole foods like fruits and whole grains higher in carbohydrates are essential for bacterial diversity. Less fiber can also add to constipation, which is a common side effect of low-carb diets, especially the keto diet. Poor gut bacteria diversity leads to weight gain.[43]

The winner, please

The ketosis diet and this fix to eat early and end early (in sync with your circadian rhythms) are both effective weight-loss methods that promote very different health-promoting habits.

An overnight version of the intermittent fasting plan (beginning early in the evening and paired with breakfast) encourages you to form positive eating habits, without adopting completely misleading ideas about which foods are "good" and "bad." This type of overnight fasting method is easy to stick to, since you are sleeping through much of the fast, and there is flexibility in the timing of your fasting regimen from day to day. In fact, avoiding food for twelve hours overnight (most of the time) throughout your life is doable and beneficial for your weight, sleep, and health, now and in the future.

Ketogenic diets, on the other hand, are short-term at best and encourage an all-or-nothing dieting mentality that promotes up-and-down weight loss and possibly greater body-fat stores in the long run.

Eating early and then ending your food intake early offers the opportunity to eat a balanced, unprocessed diet, making overnight fasting even easier to adhere to, by allowing you to feel satisfied and avoid food cravings. This, along with all of the other perks that come along with a healthier, whole-food diet, is a significant advantage over a ketogenic diet. Yet, sadly, many of the healthier whole foods are unnecessarily limited or banned on a ketosis plan.

Both plans can lead to weight loss, but overnight intermittent fasting (beginning early in the evening and ending with breakfast) is the winner for both sustainable weight maintenance and better long-term health!

BURN FAT STORES

Intermittent fasting starting early in the evening rather than later in the morning may reduce daily hunger cravings and burn more fat during the night.[44] Fasting overnight for twelve to fourteen hours (never longer than sixteen hours) triggers the body to burn its stored fat, which produces ketones for energy and results in weight loss.[45] Your body can make this switch (from using glucose for energy to burning body-fat stores to produce ketones) after just eight - twelve hours of fasting according to Dr. Steve Anton, PhD, with the University of Florida Institute on Aging.[46]

Not a breakfast eater?

The shift to eating more in the morning instead of the evening may be a challenge for some. Many adamant non-breakfast eaters tend to eat most of their food in the evening but in fact find that they actually become hungrier in the morning when they stop eating in the evening.

There is some leeway on fasting times to help with the adjustment of eating more in the morning and ending intake earlier in the evening. You don't need to eat breakfast right after you wake up, for instance. Eating breakfast two to three hours after you wake up is one way to ease into this extended overnight fasting plan. The earlier you stop eating at night, the earlier you will want food in the morning. Your body will adjust, but it may take some time to become comfortable with these eating times.

New flexible routines for fat loss

This fix for weight loss is rather easy to stick to, since people are sleeping through the majority of the fast, and there is flexibility in when you stop and start the fast. Overnight fasting permits the freedom to start fasting earlier in the evening on days when it's convenient. You can fast for twelve hours overnight some nights and longer on those evenings when it's more convenient. The all-or-nothing mentality that coincides with a more restrictive diet plan, like the ketosis diet, can be replaced with a more flexible, habit-forming approach that keeps extra weight off over time.

FIX 5

EAT EARLY

Front-load your eating. Begin a window of overnight fasting for about twelve hours most evenings, starting early in the evening and ending with breakfast in the morning.

You may have heard the saying "Eat breakfast like a king, lunch like a prince, and dinner like a pauper." That's actually great advice!

The cells in your body need time to repair and restore during the night when your digestive system is at rest. A growing body of evidence suggests that eating in conjunction with your circadian rhythm is a critical piece of the weight-loss puzzle due to the body's response to day and nighttime hormones. Big breakfasts, moderate lunches, and light dinners, which include the right types of foods, will result in more weight loss than big nighttime meals or snacks.

Time calories to eat fewer calories

Studies show that if you eat at least two-thirds of your calories before dinner, you'll consume fewer calories over the course of the whole day compared with people who eat most of their calories at night. Harvard professor Dr. Scheer reported in the *International Journal of Obesity* that dieters who ate their main meal before 3:00 p.m. lost significantly more weight than did those who ate later in the day. Moreover, the time you eat appears to be beneficial regardless of the calories, even though you may consume less, because you'll have fewer cravings.

In a 2013 study published in the same journal, two groups ate the same number of calories. Overweight women who ate a big breakfast (seven hundred calories) and a light dinner (two hundred calories) lost more than twice as much weight as did women who ate two-hundred-calorie breakfasts and seven-hundred-calorie dinners. Not only do night eaters tend to consume more calories each day, they also fail to burn more calories than those who go to sleep earlier.

Burn fat stores while sleeping

Dr. Satchidananda Panda, associate professor in the Regulatory Biology Laboratory at the Salk Institute in La Jolla, California, found in his research that you burn more stored body fat during the night while you sleep if your body doesn't have as much food to use as fuel late at night. Your body readily uses carbohydrates for fuel, and the sooner in the evening that you use up the carbohydrates for energy, the sooner you go into fat-burning mode while you're sleeping. When you time most of your eating to happen earlier in the day, you stand a better chance of burning fat stores.

Eat breakfast

Get in the habit of ending your eating before 6:00 or 7:00 p.m. so that you can go about twelve hours without eating overnight before you eat breakfast. Rather than skipping or delaying breakfast to achieve your twelve-hour fast, stop consuming food early in the evening when your body is more insulin resistant. It's important to eat breakfast relatively soon after you wake up when your insulin is most effective at doing its job of transporting glucose (blood sugar) from the blood into the cells where it can be used for energy. Insulin becomes less effective at shuttling blood sugar as the day progresses and is the least effective at night.

Not only does your body metabolize food more efficiently in the morning but eating (not drinking) a protein-rich meal for breakfast can also help you avoid evening cravings. According to research findings using brain scans, breakfast containing at least 30 grams of protein that is chewed (rather than sipped) suppresses the reward-driven, pleasure-seeking part of the brain, so you're less likely to crave less healthy, processed snacks in the late afternoon and at night.

Nighttime eating

Eating food at night conflicts with the nighttime pathways that encourage sleep. Hormones like melatonin and growth hormone are active at night and interfere with insulin's ability to transport blood sugar out of the blood. More insulin is produced to handle the higher blood-sugar level, which keeps your body from burning your fat stores.

Day, rather than evening, is when you need fuel to be the most productive. With differing work schedules, adjust these guidelines to your time frame.

Action Plan

- ✓ Start your day with a high-protein breakfast containing at least 30 grams of protein that you chew (rather than drink) to be more satisfied and to reduce your cravings for processed, high-calorie foods in the late afternoon and evening.

- ✓ Don't eat after dinner. Brush your teeth for the night, right after dinner, to avoid the temptation to snack.
- ✓ Finish eating and drinking caloric drinks between 6:00 and 7:00 p.m. to fast for twelve to fourteen hours most nights. For serious weight loss, fast up to (but not more than) sixteen hours overnight by starting your fast earlier in the evening rather than skipping breakfast, for the best results.
- ✓ Drink water freely in the evening and throughout the day.
- ✓ Fuel yourself with plenty of fruits and vegetables and healthy high-protein foods during the day (along with some healthy fats and a few servings of whole grains) to improve your energy and concentration and to help you avoid food at night when you are not as active.
- ✓ Eat a large breakfast, a moderate lunch, and a small dinner.
- ✓ Don't skip eating food in its whole form as a means of reducing your calories. You need to eat enough nutritious foods to keep your metabolism up and nutrients coming in. Do skip the processed junk food!
- ✓ Satisfy any food cravings with healthier alternatives such as herbal tea, raw nuts, or a handful of veggies. This applies to any time of the day but is especially helpful when you're trying to curb the habit of late-night snacking.
- ✓ Engage in relaxing non-food-related activities after dinner that make you feel better, such as walking the dog, reading, or enjoying a hot bath.
- ✓ Track your progress on the *12 Fixes to Healthy* app.

FIX 6
MOVE MORE

"Move More" probably doesn't mean what you think it does! I know that not everyone wants to slug it out in the gym.

Fix 6 to move more is all about how to integrate movement into YOUR lifestyle, no matter your fitness level! Moving feels good, and this chapter is all about how to do it in a way that is both scientifically sound AND can be easily integrated into your life.

CHOOSE YOUR OWN ADVENTURE

If you've never read one, the *Choose Your Own Adventure* series is a collection of books that have multiple endings. You, the reader, arrive at your own ending or destination through the choices you make.

Let's say you're hiking through the Amazon jungle. You reach a point in the story where you come to a fork in the road, and you have two choices: Do you turn left, or do you turn right? Your decision leads you to a unique result at the end of the book.

Small is big. Over the months, years, and decades, those small choices turn into habits, which in turn, have significant impacts on your weight and quality of life.

In one day, you make over two hundred food choices alone. Those are some of the forks in the road. Breakfast, or no breakfast? What kind of snack do you eat as you rush out the door? Should you go ride your bike, or should you watch a TV show? Should you go to sleep now, or keep surfing the net? You have the power to choose your journey by ensuring that your small, seemingly insignificant steps are as healthy as possible. You get to direct your own ending!

That's where exercise and sleep come into play. The amount and quality of each of these affect your food choices, along with your muscle mass, strength, and overall health, in a drastic way.

In this chapter, our fitness and wellness expert, Traci Fisher, will take you through the basics of how to make the most out of your movements, regardless of your current activity level.

MOVE MORE BY TRACI FISHER

Move more. It really is that simple. And it applies to everyone who is interested in living a healthy, strong life. "Move More" is a brilliantly simple way to summarize a potentially complicated subject. We have access to more resources than ever before. In an instant, we can access unlimited ideas on how to move, what not to do, and what we should do. It is literally endless and can be absolutely overwhelming.

The word "move" can mean many things: to pass to another place, to proceed to a certain state, and to change position or posture. Our focus for movement is not just to do it, but to do it effectively. When you hear "Move," think *quality*. When you hear "More," think about building on that quality in a way that suits your lifestyle and goals.

We will explore the quality and quantity of movement and how both can be incorporated into your daily life. When we aren't sure what to do amid the mountain of information, we can simply return to the timeless guidance of this small phrase, "Move More." Let's get moving!

Our bodies are amazing machines. They move us through life, taking care of daily activities, keeping us upright in our chairs and cars, and constantly taking in information and processing it through our brains. Most of our daily movements are auto-

> **EXERCISE ON AN OLD BRAIN**
>
> Physical activity in healthy older adults reduces the risk of cognitive decline and cognitive impairment, including Alzheimer's disease. Brain blood flow declines by about 5 percent per decade, but exercise increases blood flow and oxygen to the brain. Exercise also helps reduce blood pressure, markers of inflammation, and lipids as well as improved function of the endothelial cells lining the blood vessels.[1]

matic, and we can thank the nervous system and its ability to adapt to our ever-changing environments without active thinking. We have major systems within our bodies that work together constantly, creating movements for us. In order to have optimal movement, we need to fuel our bodies with proper nutrition and rest. We also need to fuel them with effective and purposeful movement! Yes, the ways we move and the quality of that movement both have a tremendous impact on our vitality. First, let's take a look at the power of movement and its impact on our lives.

THE "MOVE" IN MOVE MORE

Our ability to move is one of the most important aspects of our existence. As we move, our bodies and minds benefit tremendously. A cascade of neurochemicals and growth hormones are released. Movement promotes brain cell repair, lengthens attention span, boosts decision-making skills and self-esteem, improves executive functioning, and prompts the growth of new nerve cells and blood vessels. The mental and emotional benefits alone are worth the effort. Physical activity has been found to reduce cravings in alcoholics, treat depression, reduce symptoms in Alzheimer's patients, treat ADHD, and much more. We are literally wired to be rewarded chemically when we move.

> **EXERCISE TO THINK AND FEEL BETTER**
>
> Physical activity improves mood by actually changing the brain's chemistry. Exercise helps optimize the levels of nerve growth factor called brain-derived neurotrophic factor (BDNF) that is produced and secreted in the hippocampus, which is responsible for memory and learning.[2]
>
> Exercise also helps decrease stress by reducing cortisol, which leads to atrophy of the hippocampus, especially in individuals with mood disorders including depression and anxiety.[3] The bottom line—move more to think and feel better!

Movement has a multitude of physical benefits. It reduces the risk of many diseases and health conditions: cardiovascular disease, type 2 diabetes, metabolic syndrome, stroke, osteoporosis, and some cancers. We are discovering more evidence every day to prove what we already know: moving feels good mentally, emotionally, and physically.

A simple definition of movement is the changing of state, position, or place. With physical movement, we can literally see a change in position or place. Movement also includes changes in state that we cannot see but that are just as vital to our health.

We have to move to live our lives: to get from one location to the next, to complete daily activities, and to communicate. We also move to enjoy life to its fullest. We move on purpose in ways that allow us to experience adventure, enjoy one another, and create meaning. And this is just the movement we can see! There is also a lot of movement going on beneath the surface. Our bodies are constantly moving in voluntary and involuntary ways, responding to our internal and external environments. Movement is dependent on major systems within our bodies, which include muscular, skeletal, nervous, cardiovascular, and respiratory systems. Optimal movement occurs when we are able to efficiently and effectively utilize our systems across the entire spectrum of movement.

JUST A FEW BENEFITS OF MOVEMENT

- Helps manage chronic pain
- Wards off viruses
- Reduces the risk of diabetes
- Reduces cancer risk
- Strengthens muscle
- Maintains mobility
- Improves coordination
- Strengthens bones
- Oxygenates the body
- Strengthens the heart
- Clears arteries
- Detoxifies the body
- Improves complexion
- Decreases risk of heart disease

Human movement system

The health of our movements is dependent on major systems in our bodies. The Human Movement System (HMS) is made up of the muscular, nervous, and skeletal systems. The HMS utilizes these three different systems in an intricately interwoven way to create movement. The nervous system acts on the muscular system to contract muscles, the muscular system acts on the skeletal system to create movement, and the skeletal system acts as a protective case for the nervous system. This system is known as the kinetic chain.

The support system for movement is the cardiovascular and respiratory system (cardiorespiratory system). As we move, our breathing and heart rates vary, depending on how much energy we need to complete the movement. How hard our heart has to work (heart rate) to achieve a certain movement is a strong indicator of the health of our bodies and hearts. All of these systems complement, support, and protect each other. We are able to twist, turn, run, jump, and enjoy deep breaths because of these systems. They are fueled with energy through nutrition, rest, and, yes, movement!

Our bodies need to move to process fuel as well. The most basic and subtle level of movement is in our bodies' supportive systems. Digestion, respiration, sweating, dilation, and constriction of blood vessels are just a few of the types of movements that occur beneath the surface. When we move physically, we are supporting all of our systems. Movement is a source of fuel in and of itself. So, of course, we want to ensure that we keep moving into old age. One of the ways to do that is to start with a strong foundation of proper movement mechanics and then build upon it.

> **TEENAGE BRAIN AND EXERCISE**
>
> Without enough physical exercise, adolescents may be impairing their academic achievements and overall cognitive abilities. The teenage brain matures and sees an elevated production of new neurons, but stress during this time can prevent new neurons from being made. To an extent, physical activity can protect the brain from some of the effects of stress.[4]

Movement spectrum

The movement spectrum is based on both mobility and stability. Mobility is the ability to produce a desired movement, and stability is the capacity to resist an undesired movement. When we are able to stabilize our joints, ligaments, and muscles in an organized way to produce full mobility, then we have optimal movement.

Think of how your body moves on a roller coaster ride. If we didn't provide some internal resistance, we would be tossed around too much, so we "brace" ourselves to resist movement. As we brace, we are creating stability. It's not visible, but even as we walk around, we are still stabilizing some parts of our body while mobilizing others just to keep our balance. Our bodies are constantly stabilizing and mobilizing to maintain equilibrium.

This is exciting news if you don't feel ready to start an exercise routine. You can focus first on the quality of movements you already make without having to worry about increasing quantity. Sometimes getting moving can conjure up images of endless hours of running or working out in the gym, and motivation can fizzle. It is refreshing to know that just by focusing on the quality of movement, we are making a tremendous impact on our health.

Your movement

So, the focus of "Move" is quality. Quality movement occurs when our bodies' systems work together efficiently and effectively, in all types of environments, and across the entire spectrum of movement. So, how do we know if that is happening?

An easy way to determine the quality of movement is to take a look at our own stability and mobility. Stability and mobility include balance, coordination, flexibility, strength, and muscular endurance. There are many ways to test your quality of movement. The easiest is to take a look at current movement patterns as well as aches and pains. If you are in pain, your body is talking to you. Make sure that you listen! We want to use all information, including pain, to assess the quality of our movements. If we have a full range of motion at the joints, good posture, a strong center of gravity, stability during movement, and enough strength and endurance to complete the tasks that are important to us, then we are on our way to quality movement.

> **AFTER-MEAL MOVE TIP FROM JUDES**
>
> It turns out that a quick ten-minute walk after a meal is especially good for you! Your body handles the sugar from the breakdown of your food the very best in the morning. As the day goes on, insulin is less effective at moving sugar out of the blood and into the cells for energy. Elevated blood sugar levels lingering too long often put your body in fat-making mode.
>
> A short walk (or any movement after a meal), for even just ten minutes, helps control your blood-sugar levels. Your muscles use this extra blood sugar as energy, so there is less circulating in the blood. A walk after your evening meal is particularly good at controlling those blood glucose levels that may rise too high and stay elevated too long in the evening, especially as you age.[5]
>
> Faster digestion with lower rates of heartburn and reflux symptoms are bonus benefits from an after-dinner walk. Part of your Move More plan may translate into an extra bit of movement after dinner before you hit the couch for some screen or reading time.

There are many ways to improve the quality of your movement. Focus first on improving the movements that are closest to the center of your body. Think of your lower back (stability), your pelvis (mobility), and your upper back, shoulder, and chest (thoracic mobility). Your shoulder blades need to be stable to support all the movement at your shoulder. Once your body is stable in the right spots, you can teach your body to move and balance as a whole. The fastest way to improve is to practice whole movements with proper form. If there is a movement that you cannot perform, practice that exact movement in a controlled environment with support like a chair. Then, slowly, as your body learns, take away that support. Remember, quality first, then quantity!

THE "MORE" IN MOVE MORE

If we want to counteract the process of aging and keep the benefits of movement, then we need to strive for more. Why? Because we have amazing bodies that thrive on being challenged. This is called specific adaptation to imposed demand, which means that the human body is designed to adapt to any stress placed on it. Over time, our bodies learn, and then certain movements no longer take as much energy, either mentally or physically.

Why more?

The body's ability to learn and adapt has served us well across the years. We used them to hunt, farm, build homes, and move in many ways to protect ourselves and to stay alive. We evolved, and we are still evolving, just in different ways. Now, we still need to Move to survive. Because our bodies are built to adapt, that is exactly what they do.

As we spend more time hunched over our desks and less time upright and moving, our bodies adapt accordingly.

Some muscles work overtime, while others are underworked. This can cause aches or injury and eventually leads to muscle imbalances and a host of other physical issues.

The positive news is that the body's ability to adapt to stress never goes away. As you change and challenge yourself, your body becomes more efficient and stronger, and the benefits of moving add up. Better yet, this includes any change at all—even ones we think are insignificant. That's why we want you to Move, and to Move More!

FITT

"More" refers to the type of movement, the intensity of movement, and the amount of time spent moving both in terms of frequency (number of times per week) and the amount of time spent on the activity each day. An easy way to remember how to add More is by thinking FITT: Frequency, Intensity, Time, and Type. Moving our bodies should also be fun and make us feel alive.

Whether you are working on the quality of your movement, increasing the quantity, or both, your body begins to change whether you see it or not. Keep in mind that just by improving the quality of your movements, and therefore increasing balance, stability, and mobility, you are already adding More.

The American College of Sports Medicine (ACSM), Centers for Disease Control and Prevention, and American Heart Association all have recommendations regarding the frequency, intensity, time, and type of movement that will bring optimal health. The recommendations vary based on age and current levels of fitness, but in general, they recommend at least either 150 minutes of regular activity OR 60–75 minutes of vigorous activity per week. They also recommend at least two days per week of strength and endurance activities. Keep in mind that these guidelines are goals for basic health. Start where you are, and then progress from there.

TYPE OF MOVEMENT

Twist, turn, dance, sprint, bend, reach, jump, spin, sit, contract, lift, exhale, stabilize. Whatever type of movement you choose to engage in, first and foremost, it should have quality. We said that we can define quality movement through the variables of mobility and stability. We asked, "How are you moving?" Many people may suspect that they need more balance, coordination, endurance, strength, or flexibility. Take a look at your physical history. If you have common injuries or aches, focus on stability and mobility first.

> **PREGNANT MOVEMENT ON BRAIN DEVELOPMENT**
>
> Pregnant women who are physically active before and during pregnancy improve their babies' intellectual development and performance. Studies report a positive relationship between maternal exercise and improved reading comprehension, scores on intelligence and learning tests, as well as attention levels in school-age kids.[6]

TIME AND FREQUENCY OF MOVEMENT

You can increase the quantity of your movement through either more time per activity or greater frequency or both. For example, you can set a timer while working at your desk to remind yourself to stand up every hour for a minute. Or if it suits you better, stand up every ninety minutes for several minutes. The old boring suggestions, like parking a little farther away or taking the stairs, actually do make a difference!

Improving the quality of your movements (range of motion, balance, and some muscular imbalances) actually takes no extra time at all. For example, if you have already figured out that you need a bit more balance in your life, try standing on one foot while you engage in everyday activities like brushing your teeth, standing in line, or doing the dishes. You can also use this time to check out your range of motion just by literally moving your body at its joints (arm circles, neck rotations, hip circles); the options really are endless!

Perhaps you are spending too much time hunched over at your desk, and your shoulders are starting to bend a little too far forward. Every time you get up from your desk, reach back behind you, clasp your hands together and lift your chin toward the sky.

DON'T HAVE TIME?

Here are some quick tips you can use to "Move" a little "More" than you did yesterday:

- Squeeze your butt muscles 3–5 times at every red light.
- Don't sit on the toilet; squat over it.
- Hold groceries, your cell phone, purse, and dog leash with the opposite arm than typically used.
- Stand on one foot while brushing your teeth.
- Park farther away than you did yesterday.
- Take the stairs at work, at the hotel, at the mall . . . everywhere!
- Every time you answer the phone, stand up and walk around.
- Set a timer to stand up and move around at least five minutes per day.
- Breathe very deeply as often as possible!

INTENSITY OF MOVEMENT

Don't let the word "intensity" scare you. Intensity is progressive, and it starts wherever you are. Increasing the intensity of an activity can mean doing more of it (frequency, time, type) or increasing the exertion or effort that you are expending. We have all seen the gadgets available to help us measure our intensity. But don't forget that your brain and heart are useful tools too. As you begin to exert more effort, your body knows it.

You know if a box is getting heavy or if the light bulb you are replacing above you is starting to feel like it has quadrupled its weight. This is called your rate of perceived exertion. Sometimes this happens naturally, and sometimes we make it happen on purpose.

For cardiovascular endurance, you can obviously increase the intensity by doing a movement more often or faster. We can also increase the intensity by moving on different planes of motion. If you like to speed walk, turn sideways for part of the walk. Fortunately, it's easy to increase the intensity, and we can do it pretty quickly.

You define what "More" means to you. We all have different circumstances, capabilities, time, and bodies. For some, moving more is standing up several times throughout the day at their desks. For others, it's completing a physical therapy exercise or getting into a strength-training program. It could mean something completely different to you. The bottom line is to take a look at how you are currently moving and see how you might add "More" in a way that supports what you need.

PLANNING WITH JUDES

Make plans B and C and even D when you decide your Move More goals. When the crazy days or just plain exhausted days roll around (and they will), you'll be ready! Your plan A exercise goals may be lofty, but your plan D may be a walk around the block or a ten-minute exercise routine while watching Netflix. What matters is that they are part of the plan, ready in your mind. Besides, plan D may be humble, but you will feel better, and a little movement over time can make a big difference: small is BIG!

This brings us to your Move More formula. Personalize it by thinking about YOUR ideal weekly movements. Now think about your current movements in terms of their quality and quantity.

YOUR MOVE MORE FORMULA

This table combines Move and More across the entire spectrum of movement. Use it to increase your awareness of your current movements and how you might add a little More to get to your ideal movement pattern.

	FREQUENCY	TIME	INTENSITY	FUN!
	How many days / week?	*How long do you spend?*	*How hard are you working?*	*Do you enjoy this movement?*
STABILITY *Balance / Posture*				
MOBILITY *Range of Motion / Flexibility*				
STRENGTH				
CARDIOVASCULAR				

FUEL TO MOVE BY JUDES

I've found that fueling for a game of tennis makes a huge difference to my energy level on the court. Refined carbs cause my body to drag in a shorter time while whole grains give me lasting energy. I can tell when my body and brain don't have enough fluid too. My decision and movement reaction times are slower. I can feel the contrast in my drive and stamina, and so can my opponents!

Nutrition is a critical part of a good workout. You need to give yourself the necessary fuel, hydration, and electrolyte balance to sustain high energy levels and build strength when you're exercising. If your body is an engine, foods are what you use to fill up the tank, and fluids for the radiator. Without proper fuel and fluid, you're likely to burn out much faster.

There are certain amounts and kinds of protein, carbohydrates, fluids, and electrolytes you can consume to maximize your workouts. These vary depending on the type of activity you're doing, for how long you're doing it, and the intensity of the exercise. If you're a competitive athlete, see a registered dietitian nutritionist (RDN) that specializes in sports nutrition for help to become a better competitor. Below are some general guidelines for those participating in less competitive, recreational, or exercise-related fitness.

Before workout

If you exercise first thing in the morning, dehydration is your biggest concern. You've most likely gone the night without a sip of water. Because all your cells rely on fluid to function, your workout, mood, and even thinking skills will suffer if you're not drinking enough water!

Your next priority is to fuel your engine! Your body can't run on fumes for long. If you haven't eaten all night, at least have a little snack of yogurt or bananas with peanut butter before you exercise. These carbohydrate foods are converted into fuel for your body. Then back up your workout to a meal within a couple of hours after ending your exercising.

Steer clear of overeating right before you work out. Stop eating complex carbs two to three hours before your workout. Thirty minutes before working out, eat a small, simple carbohydrate like yogurt, a banana, or an apple. Simple carbs in whole foods like fruit and dairy are a better choice than simple carbs from sugar. The thousands of nutrients in fruit and yogurt help keep the fuel to your muscles and brain steady. Sugar found in processed foods often causes your blood glucose level to crash sooner.

Your primary focus before exercising should be on taking in carbohydrates for energy and fluids to hydrate. Protein alongside carbohydrate-rich foods helps to steady that fuel level and to repair muscle damage. If you're participating in resistance training like lifting weights, then pre-exercise protein can help reduce the breakdown of the muscles that occurs during your workouts. Protein before resistance exercise doesn't translate into bigger muscles, however. Eat real food containing carbs and protein, along with water, instead of relying on protein bars and shakes. Remember, food in its whole-food form offers so much more than processed foods, as thousands of nutrients work together synergistically to benefit your health.

Workouts less than an hour

For moderately intense workouts lasting an hour or less, you shouldn't need to eat any extra carbohydrates as long as you ate a whole grain or fruits and vegetables earlier in the day. For morning workouts, drink 8 to 10 ounces of water and eat a half or whole banana with peanut butter, Greek yogurt, or a half piece of whole-grain toast and an egg before you begin. Then eat the rest of your breakfast when you finish the workout. Even a few bites of your breakfast before you exercise helps you have a more effective and enjoyable workout.

Workouts more than an hour

For intense exercise sessions lasting more than an hour, fuel your body with a mixture of whole grains and simple whole-food carbohydrates. A piece of whole-grain toast topped with a little peanut butter or an egg along with fruit will do the trick.

Drink water throughout the day of your workout. One or two hours before the workout itself, drink 15 to 20 ounces of water. Fifteen minutes before you begin exercising, drink 8 to 10 ounces of water.

Intense workouts several hours

Pasta seems to be the typical food that comes to mind when you think about carb loading. For workouts lasting several hours, such as marathons or triathlons, loading up on carbs is a good way to make sure that energy stores are filled to capacity. However, it's best if you load up on complex carbohydrates starting a few days before the race or game. Complex carbohydrates include whole grains (like whole-grain pasta instead of pasta made with white, refined flour) but also include starchy vegetables, beans, and lentils, providing long-lasting fuel for exercise plus so many nutrients and fiber.

Eat a low-fat, low-fiber, low-protein, high-carbohydrate meal three to four hours before the game. Note that I'm giving you permission to eat more refined carbs for workouts lasting several hours. Eating carbohydrates before intense training increases endurance by about 20 percent. Eating whole grains just before an event may be harder to digest while exercising.

Hydration is also critical to be able to think clearly and perform well. Drink plenty of fluids with meals, as well as about 16 ounces of water (5 to 10 mL/kg body weight) two to four hours before the event. Leave enough time to void excess fluids, and make sure your urine is a pale color.[7]

During workout

Drink water frequently while you're exercising. As a rule of thumb, drink between 13 and 26 ounces of water during each hour of exercise. However, if you are sweating and you've been working out for more than one hour (maybe sooner if you are sweating profusely), it's important to replace the electrolytes and energy stores you've lost during your exercise. Drinking water can dilute the sodium in your blood and make your electrolyte situation worse—especially if you are sweating a lot.

Endurance sports drinks (without artificial colors) can usually replace the sodium and potassium you need, along with fluids. Eating a salty snack along with fluids along with an endurance drink isn't a bad idea just to make sure you have enough sodium. The drink should contain about 110 milligrams of sodium and 30 milligrams of potassium in 8 ounces. A sports drink with 14 to 15 grams of carbohydrates per 8 ounces can help replenish the energy stored in your muscles. A lighter version sports drink with fewer carbs and calories works too if your workout isn't too long.

Beyond sports drinks, bananas are another excellent source of sodium, potassium, and carbohydrate for energy. You can eat part or all of a banana whenever you stop to rehydrate.

After workout

Hydration. For short bouts of exercise, drink 1.5–2.5 additional cups of water after you finish. For longer workouts, weigh yourself before and after you exercise and shower. Drink about 16 ounces (half a liter) of extra fluid for every pound lost through exercise. Some of that fluid should contain carbohydrates and electrolytes (like milk), or you can consume food with your water to replenish your electrolytes and the energy stored in your muscles. Be generous with the salt shaker after a long, sweaty workout.[8]

CHOCOLATE MILK

Contrary to popular belief, the extra sugar in chocolate milk is necessary only if you need to refill your energy stores for another intense physical workout that same day or the next, such as for an athlete who has multiple events in a day or has a strenuous training schedule. Otherwise, regular milk and other dairy products contain enough carbohydrates to refill your energy stores without the extra sugar in chocolate milk. Regular meals and snacks after your workout usually provide more than enough carbs to replenish your muscle energy stores.

Carbohydrates. Dairy foods are also good at replenishing carbohydrate stores. For workouts lasting more than an hour, you need to replace the carbohydrate stores in your muscles. Start restoring them with regular meals and snacks within the first thirty minutes of your extended workout.

Protein. Your muscles need protein to recover and build. Traditionally, the best time to deliver that protein is within thirty minutes up to two hours after exercising. However, newer research indicates that our muscles continue to build for twenty-four hours after a workout more effectively than once thought.[9] More recent recommendations for protein suggest regular spacing of moderate amounts of protein throughout the day rather than depending on protein supplements or power bars immediately following a workout.[10]

But keep in mind that more isn't better. The body uses only about 20 grams of protein right after an exercise session for maximum muscle-building benefits. Most protein shakes are way too high in protein, and whole foods like Greek yogurt can easily replace protein shakes. Or make your own shake at home using Greek yogurt and frozen fruit or peanut butter!

Protein supplement alert

Leading exercise physiologist Dr. Blake Rasmussen from the University of Texas explains that as much as supplement companies don't want to hear it, eating plenty of protein evenly throughout the day appears to improve muscle building without the need for protein supplements right after a workout. That can vary in competitive athletics, but for most people involved in a fitness-related exercise to improve their health, eat 75 to 150 grams of protein spread out evenly between meals and snacks.[11] Does that recommendation sound familiar? That's Fix 2 of the *12 Fixes to Healthy* plan in this book!

More than 25 grams of protein immediately after a workout doesn't help build more muscle, but the type of protein does matter.[12] Protein high in the amino acid leucine appears to be an excellent way to boost muscle building in the short term after intense physical activity. Not only are dairy products such as milk, cottage cheese, and yogurt the richest source of this muscle-building amino acid, but they also contain carbohydrates to replenish your energy stores, electrolytes to restore your electrolyte balance, and significant calcium to protect your bones. Peanuts, beans, lentils, soybeans (edamame), poultry, seafood, pork, and beef also have good leucine content. You just need more of them to reach the levels of leucine found in dairy food like cottage cheese and yogurt.

Choose whatever is most comfortable for you, but try to eat and drink this real food as opposed to processed protein shakes and bars when possible.

FIX 6

MOVE MORE

Move your body more.

The "Move" in Move More ultimately means quality movement. Quality movement is the foundation for all other movements and is based on stability and mobility. That means that your body can stabilize itself in some ways while at the same time mobilizing in other ways. Think of the balance required on a step ladder as you rotate your shoulder to twist in a light bulb. You need both mobility and stability to do this safely.

Another way to see the quality of your movement is to pay attention to your posture, balance, and ability to have a full range of motion at all your joints. If you know that you are limited or have recurring aches and pains, you need to address them first. It doesn't take any extra time to incorporate proper movement into the movements you already make every day. Try moving in the opposite way than you normally do. This can mean looking up toward the sky before looking down at your phone or contracting your back muscles at your desk. Rotate your wrists in circles for thirty seconds every hour if your joints tend to ache. Become aware of how you move every day, and make sure that you focus on adding quality to your current movements before you add quantity.

Quantity is where the "More" of Move More comes in. A way to do this is by applying the FITT (Frequency, Intensity, Time, Type) principle to your movement routine. Changing up any one of these variables means adding "More" to your "Move." Think about how you are currently moving. If there are any challenges to your balance, posture, or stability, incorporate an activity that will strengthen those areas first. Just by focusing on the quality of movements, you'll already be doing "More" by increasing the variables of type and frequency. Adding intensity and time can mean speeding up, moving in different directions, adding more weight, working out for longer—the options are endless.

"Move More" is simple, but that does not mean it is easy. As with all the fixes in this book, our internal and external environments affect our choices. Here are some quick tips on how to make "Moving More" a reality you can build on—starting right where you are.

Make it fun, and make it yours! Moving our bodies should feel good. Discover how you like to "Move," whether it is gardening, golfing, walking around the mall with a friend, sightseeing, or discovering new group fitness classes in your area. Choose someone you want to strengthen your relationship with, and ask them to join you.

Use the *12 Fixes to Healthy* app to share your progress. Fit movement into your life instead of trying to fit your life into a program.

Partner up! Accountability may sound intimidating, but it works. Find someone, like family members, friends, or coworkers. If you prefer to work out alone but still want accountability, find someone you don't know in an online program.

Start small! Yes, all those little tips about parking farther away from the building, taking the stairs, and holding groceries instead of using a cart do add up. Be creative and clever and see how you can move more by choosing different habits. (Stop parking the car at your mailbox to get the mail!)

Track it! The number one tip that I have for increasing fitness levels is to track your activity. Track using pen and paper or the *12 Fixes to Healthy* app. This simple act of tracking automatically creates awareness and results.

You now have a simple way to think about and break down what it means to Move More. Regardless of your current activity level, age, or circumstances, remember that your body is the most amazing machine on this earth. Take a fresh look at how you move. As long as you are choosing a variety of movements that work on strength, cardiovascular, coordination, balance, and flexibility, you can define your own ideal movement patterns. Keep various ideas in mind that suit your mood and time constraints.

Remember, rigorous exercise is optional, but regular movement is mandatory. Keep challenging your body. Create the adventure you inherently deserve!

Action plan

- ✓ Choose new ways of moving to incorporate into your daily life. Take the stairs instead of the elevator, stand on one foot, stretch every time you stand up.
- ✓ Think of a type of movement that you really enjoy like dancing, tennis, or gardening, and make time for it!

- ✓ Use the Movement Formula Table to see where you can add in a little more frequency, intensity, time, or type of movement!
- ✓ Whatever exercise you do, write down the specifics at least one time. One month later, do it again and see how much you have improved.
- ✓ Check your current workout plan to see if you are hitting all the elements of a well-rounded program: strength, cardiovascular, coordination, balance, and flexibility. If you're not, replace one session with another type of training.
- ✓ If you are rushed or less motivated on any given day, keep in mind a shorter, easier exercise plan as a backup plan rather than forgoing all exercise that day. It might be as easy as walking, getting up from your desk, or even doing deep breathing muscle relaxation exercises!
- ✓ Regardless of your current activity level, check the quality of your movement to ensure that your body is in alignment.
- ✓ Fuel and hydrate your body before you exercise for better success and more enjoyment.
- ✓ Track your progress on the *12 Fixes to Healthy* app.

FIX 7
WATER WELL

Water is necessary to burn body fat. Does that give you more motivation to take your water bottle with you wherever you go and keep it in arm's reach?

Will you replace caloric drinks with water more often if you know that caloric drinks put you in a fat-making mode more quickly and for longer than even caloric processed treats?

Water plays a critical role in your digestion, absorption, circulation, creation of saliva, transportation of nutrients, and the maintenance of your body temperature. About 60 percent of your body is made up of water. So don't miss out on drinking this important, life-sustaining nutrient like the 43 percent of adults who drink fewer than four cups of water a day.

Follow Fix 7 to drink water throughout the day and evening instead of other beverages!

HYDRATE TO BURN FAT

Drink up, everyone. Water can help you lose fat weight and keep it off. Exciting research has shed light on how drinking water helps to burn body fat. Cells in our body need to be swollen with water to burn body fat more efficiently. When our cells are low in water, they store more fat instead of using it for energy. Of course, once the cells are completely hydrated, more fat won't be burned by drinking more water. But drinking water consistently ensures that the fat cells can burn to their full capacity.

Without water, the body can't properly metabolize body fat, a process called lipolysis. The first step of this process requires water. Drinking enough water is essential for burning body fat, as well as fat in food. A 2016 review of the scientific research on hydration and weight found that mild, ongoing dehydration fosters obesity while increased water consumption leads to increased body fat loss in animals.[1]

When your body is dehydrated, it cannot burn fat. So drinking water (not soda, alcohol, or energy drinks) is critical to burning your fat stores! As a bonus, your skin will glow and your hair will shine.

HYDRATING FOODS

Water in food counts too! Fruits are full of water. Watermelon ranks highest on the list being 90 percent water. Melons like cantaloupe and honeydew as well as oranges and grapefruit are also water bombs.

Vegetables may not contain the water percentage of fruit, but their nutrient-rich water source is nothing to scoff at. Celery, cucumbers, bell peppers, tomatoes, and Romaine lettuce are among the highest water contenders. Soup and smoothies may be an obvious water source, but yogurt and oatmeal are less obvious. Reach for hydrating foods to keep your fat cells swollen with water and burning fat more efficiently.

Caloric drinks

Drink water instead of caloric beverages during your meals and snacks so that you can burn the calories you are eating more readily. Drinks with calories produce too much insulin too quickly and linger for too long after the meal. This prevents fat cells from burning as much energy.[2]

Dr. Jodi Stookey at Children's Hospital in Oakland even recommends replacing milk with yogurt, pointing out that milk was one of the caloric beverages used in the studies revealing water's advantage in weight loss.

The advantages of water don't stop there. You receive the extra benefit of skipping high-glycemic calories when you skip caloric drinks. In fact, research has shown that people don't eat additional food to compensate for calories saved from drinking water instead of caloric beverages (although the same can't be said about diet sodas).[3]

Decreasing even three hundred calories every day makes a big difference over time. Given that caloric beverages are mostly simple sugars, avoiding these calories is even more powerful in reducing your fat weight. We often confuse our bodies' signals for thirst as a sign to eat.[4] Sometimes, our mouths just need something to do. Water fills you up, and it doesn't add on calories the way that munching on snacks or drinking caloric beverages does.

Burn more calories

Drinking water can help you lose weight by revving up your metabolism. A study published in the *Journal of Clinical Endocrinology and Metabolism* found evidence that water speeds up the rate at which you burn calories by as much as 30 percent within just thirty minutes of drinking it.[5] That may not sound like a lot of extra calories burned, but it adds up over time, and it is especially significant when water is consumed along with a meal (or right before or after a meal).

Drink to lose

Overweight women who increased their water intake to over 1 liter (34 oz) per day lost an extra 4.4 pounds over one year. These women didn't make any other lifestyle changes except to drink at least a liter of water more per day.[6] Drinking water is an easy fix to a difficult problem and may have other bonus perks.

WEIGHT LOSS WATER TIP

Drinking 16 ounces of water, three times a day, before your meals may help you lose weight. A controlled study that compared those who drank a pint of water thirty minutes prior to eating their three meals with those who didn't, found weight loss was greater with the pre-meal water drinkers. Participants who loaded up on water before meals lost an average of 9.48 pounds, while the other group lost an average of 1.76 pounds over twelve weeks.[7] Try drinking 16 ounces of water before your meals and see if this simple trick helps you lose weight.

HYDRATE TO THINK

When you're low on fluids, your brain triggers thirst, meaning your body is already mildly dehydrated by 1 to 2 percent. Recent science has indicated that your thinking or cognitive abilities can be impaired even at this mild level of dehydration.[8] When you get those cues, you need to listen to them so you can learn, recall, and make better decisions. As a bonus, you may be in a better mood as well. While evaluating hydration and brain function, researchers also found mood improved in both men and women participants.[9] Water is a nutrient, so think of it as one. Your body loses a lot of water each day through sweat, breathing, urine, and stool. That means we need to replace it to run our bodies and our brains.

HOW MUCH WATER?

To check periodically how you're doing in terms of water intake, take a look at your urine. It should be clear to a very faint yellow color. Douglas Kalman, PhD, RD, a professor of sports nutrition at Florida International University in Miami, recommends that you divide your weight in pounds in half to estimate the number of ounces of water you need to drink per day (150 lbs. ÷ 2 = 75 oz. of water).[10] But keep those cells hydrated all the time by drinking water throughout the day rather than at one sitting.

FROM SODA TO WATER

Those trying to drop weight often believe that low-calorie or diet drinks can help them lose weight. Women drank water instead of a diet drink after their main meal midday over a twenty-four-week diet program in a study published in the journal *Diabetes, Obesity and Metabolism* to test this theory. The water-drinking group lost on average 2.5 lb. more than the women who drank diet drinks and improved their insulin sensitivity.

The researchers reported that by replacing diet drinks with water, the women adhered better to the weight-loss diet. This study indicates that while they can still lose weight, they may not be losing as much as they would if they drank water in place of diet drinks. The results of this study also questions whether consuming diet drinks is the best way for people with diabetes to manage their glucose levels since the diet soda drinkers crave sweeter foods over the water drinkers.[11]

There also seems to be some evidence that diet sodas alter your gut microbiome balance to promote weight gain and type 2 diabetes.[12] More research is needed to solidify this idea, but it appears to have grounds for interesting investigation.

Diet or regular, if you're an avid soda drinker, water may seem boring. Flavor, fizz, and caffeine give soda the edge over water. There are ways to spruce up your water to make the transition from soda to water easier and more interesting!

Flavored water

Add some pizzazz to your water by lightly flavoring it. Try sliced cucumber or wedges of lemon, orange, or lime. Frozen fruit like berries or pineapple keep your water cold and sweeten your water, all at the same time. Fresh mint, ginger, or basil adds a refreshing zing to your water. Find a combination you like.

How about tea instead of soda? If it is flavor you're after, sipping tea, cold or hot, can be another alternative to water. Tea is calorie-free, with or without caffeine, and contains powerful phytochemicals to boost your health. Great-tasting herbal teas along with green and black teas can make the transition from soda to water easier. Fruit herbal teas make tasty, hydrating cold drinks. The caffeine in the tea may be helpful for those who are transitioning from caffeinated soda. If you're cutting back on caffeine, go for all the herbal teas available and caffeine-free tea options.

With so many different flavor options, it'll be easier to replace those sodas with water!

INFUSED WATER IDEAS

In a large pitcher, try the following combinations in about 5 or 6 cups of water. Add ice and fill the container with water. Chill and enjoy it. Note that the flavors get stronger as the infused water sits. Find out which combinations float your boat!

- Raspberries, Cucumber, Lemon
- Watermelon, Kiwi, Lime
- Honeydew, Cucumber, Mint
- Blackberries or Blueberries, Orange
- Pineapple, Lime, Coconut
- Grapefruit, Pomegranate, Mint
- Mango, Raspberry, Ginger
- Strawberry, Lemon, Basil

Dietition Ruth says her favorite infused water combo is made with fresh mint leaves, a couple of half-inch chunks of fresh, peeled ginger, and sliced cucumber. This infused water is so flavorful! You can even refill the water for a fresh pitcher each day for three days using the same mint, cucumbers, and ginger. Cheers!

Ruth Ranks MEd, RDN, CSOWM

Sparkling water

Carbonated water, seltzer, and soda water are all terms used for sparkling water. So no matter what you call it, they all have some fizz. Club soda is water-based as well, but it comes along with other minerals (including a lot of sodium) that change the taste slightly, unlike seltzer water.

Seltzer water is as hydrating as regular water and can be an excellent chemical-free, sugar-free, and calorie-free alternative to plain water. The naturally flavored seltzers may not taste as sweet as diet or regular soda, but in time your taste buds adjust. For a little extra natural flavor and nutrition, squeeze some fresh lemon or lime in your seltzer.

Seltzer is acidic compared to plain alkaline water. Teeth fare better in a neutral or even slightly alkaline solution like plain water. Over time, that acidity can lead to enamel erosion to your teeth. If drinking seltzer water on a regular basis, drink it with meals and be sure to finish with plain water. Otherwise, use a straw to allow the seltzer water to bypass your teeth.

For those transitioning from diet or regular soda to water, try drinking seltzer water with juice. It hits the sweet spot along with the carbonated rush. Gradually remove the fruit juice and opt for the seltzer alone. Seltzer brands like La Croix, with no calories, artificial sweeteners, or sodium are not too fizzy or too sweet and are an option to replace soda, avoid fat-promoting liquid calories, and hydrate your cells.

TIPS TO TRANSITION

- Before drinking a glass of soda, drink a big glass of ice water first.
- Dilute your soda with water. Try drinking half-soda, half-water.
- Flavor your water with fruit, vegetables, herbs, and spices.
- Replace soda with sparkling water.
- Try adding fizz to your water at home with a do-it-yourself fizz kit for around $75.
- Go caffeine-free first. Gradually decrease the number of caffeinated drinks so you don't have to kick both the soda habit and the caffeine habit at once.
- Switch to caffeine-free sodas as part of your transition or use caffeinated tea as your drink choice to start.
- Keep a chilled water pitcher in the fridge or chilled water dispenser easily accessible.
- Make ice cubes with fruit, vegetables and/or herbs in an ice cube tray to add to your water bottle later. - Allison Harrell, RDN, CCMS, Virtual Dietitian at www.mywholenewlife.com

ENERGY DRINKS

While energy drinks are promoted as dietary supplements that boost mental alertness, increase energy, and improve physical performance, they certainly aren't multivitamins! Some are marketed as ordinary soft drinks, which they are not. A growing body of scientific evidence shows that energy drinks can cause serious health issues, especially in kids, teens, and young adults.[13] Yet energy drinks are the most popular "dietary supplement" consumed by teens and young adults in the United States. Men aged 18 to 34 years drink these beverages the most, and about one-third of teens aged 12 to 17 years old drink them on a regular basis.[14]

There is no legal requirement to disclose the amount of caffeine on the label, and the amount varies greatly. Sometimes it's difficult to determine the amount of caffeine, and a common ingredient, guarana, adds more caffeine to the drink that is not accounted for.

Large amounts of caffeine can cause serious heart and blood vessel issues like increases in heart rate and blood pressure as well as heart rhythm problems. Caffeine also can negatively affect children's developing nervous and cardiovascular systems. Caffeine is associated with sleep problems, anxiety, dehydration, and digestive issues.

Last but not least, a 16 ounce energy drink can contain between 54 to 62 grams of added sugar or 13 to 15 teaspoons of sugar, exceeding the amount of added sugars recommended for an entire day. That brings another set of problems for our young people.[15]

> **HEADACHE WARNING**
>
> You may not realize that a headache could be caused by lack of water intake. You might be really busy running errands or working on a project, and water isn't there to drink. You're so occupied with your activities or work that you end up not drinking enough water.
>
> It turns out that there is something called dehydration headaches that occur simply because you didn't drink enough water.[16] The good news is that a glass of water is your best medicine. If you have been sweating a lot, eat a salty snack too.

KOMBUCHA

A bubbly kombucha lower in sugar makes a good swap for alcohol and soda for adults. But if you're drinking lots of kombucha as a means to improve your gut health, there are better ways.

First and foremost, consuming a variety of fiber from fruits, vegetables (including lentils, beans, and dried peas), nuts, and seeds is the best way to promote an environment for the microbes to flourish. In fact, these whole foods feed the good bacteria already in your gut so your body is able to grow the microbes you need.

Consuming extra beneficial bacteria when the conditions in your gut are poor is like planting seeds in poor soil. You're better off nourishing the good microbes already growing in your gut with a variety of plant-based whole foods full of bacteria-feeding fiber.

Microbes strength

Kombucha doesn't have nearly as many health-promoting microbes as nutrient-dense yogurt or kefir (both milk and water). There are at least five hundred different species of bacteria, which translates into about three hundred trillion microorganisms living in your gut. Even concentrated probiotic capsules contain only about five species and about fifty billion microbes. Kombucha contains far less.[17]

Kombucha risks

People may not realize that by drinking kombucha there is a risk of contamination if conditions when making the kombucha weren't completely sterile. The sugar-rich solution used to make kombucha helps not only the good bacteria to grow but also harmful pathogens (like E. coli) if the conditions aren't just right. Fungal and bacterial contaminants are not uncommon in homemade versions of kombucha and can cause mild side effects like stomach problems, nausea, yeast infections, and allergic reactions. Severe side effects are rare. Molds like Aspergillus have also been found in kombucha, and twenty cases of cutaneous anthrax with allergic reactions and liver damage were reported with homemade kombucha. Excessive intake of kombucha can also result in metabolic acidosis.[18]

Children and the elderly or anyone whose immune system is more fragile are at greater risk of getting sick from contaminated kombucha.[19]

Alcohol in kombucha

Too often, well-meaning parents push their children to drink kombucha as a means of improving their health without realizing that there is alcohol in it. Kombucha is made by adding live yeast, bacteria, and sugar to tea and leaving it to ferment for a few weeks until it turns into a slightly sweet, slightly tart beverage. Through the fermentation process, kombucha tea is left with alcohol, carbon dioxide, acetic acid (and other acids), and probiotic bacteria.

The remaining alcohol content is low, but there is a lot of variation in kombucha, and you still might be left with more than 0.5 percent, legally making it an alcoholic beverage. Tests of various raw brands of kombucha found alcohol levels up to 3 percent. The alcohol level in a can of Coors Light is 4.2 percent, a typical beer is about 5 percent, and a glass of red wine contains from 11 to 14 percent alcohol.

Even kombucha left with 0.3 percent alcohol can be a problem for people sensitive to alcohol, children, pregnant and lactating women, and those taking medications that interact with alcohol.[20]

Caffeine

Since kombucha is made from tea, it usually contains some amount of caffeine. The fermentation process does cut down on the caffeine content, but about one-third of the caffeine remains, which is about 10 to 25 mg per serving (half a bottle) for black tea kombucha. It's not an exact percentage, but be aware that kombucha does have caffeine in it. If you are drinking a whole bottle of kombucha (two servings), you're taking in a little less than a quarter of the amount of caffeine in an 8-ounce cup of coffee. If you're someone who drinks lots of kombucha in a day, the caffeine adds up.

Sugar

The process to make fermented tea requires some sugar. During the process, good bacteria feed on most of the sugar to make lactic acid. So, the end product has far less sugar than you start with, but it's difficult to know exactly how much. Tests have revealed that labels provided a much lower sugar amount than the sugar actually in the final product.[21] Check the label for the sugar amount, but know that it may not be correct. Buy a kombucha with less than 4 grams of sugar per serving.

Kombucha bottom line

Because there is a risk of contamination and kombucha contains both alcohol and caffeine, it is not recommended for children. Eat plenty of plant-based foods in their whole form to nourish your gut to help the microbes found in fermented foods provide greater benefit. Eat a variety of foods with live active bacteria including yogurt and kefir to provide a diversity of microbes and nutrients, enhancing your gut health.

Avoid kombucha with more than 4 grams of sugar per serving. Stick to one serving, which is usually half of a bottle, instead of drinking the whole bottle. Drink kombucha as a substitute for an alcoholic drink or a soda.

KEFIR

Water kefir is a fermented beverage like kombucha, but it begins with water rather than tea. It's bubbly but less sweet and sour than the pungent apple cider vinegar taste of fizzy kombucha. Although both of these fermented drinks contain healthful bacteria, kefir is a much better source of lactic acid bacteria. In fact, kefir grains can have up to fifty-six different yeast and bacterial strains, leaving kefir more potent and encouraging a healthier gut environment than kombucha.[22]

Unlike kombucha, water kefir has zero caffeine and negligible alcohol content. Similar to kombucha, kefir water does use sugar to feed the microbes. Typical homemade water kefir contains about 3 grams of sugar. Fruit juice added to your fermentation brings the sugar content up to about 5 grams for every 8 ounces. Commercially sold Kevita brand of water kefir, for example, contains 4 grams of sugar in 8 ounces. Compare that with the 25 grams of sugar in the same amount of Coke.

The good news is that contamination doesn't seem to be an issue with kefir like it is with kombucha. Water kefir grains used to make water kefir tend to inhibit harmful pathogens.[23] The water kefir grains may even be able to lower the risk of contaminated water. The grains seem to have a natural ability to absorb toxic metals as they grow.

> **WATER KEFIR WINS OVER KOMBUCHA**
>
> With superior probiotic strength, no caffeine, less alcohol, and less risk of contamination than kombucha, water kefir wins over kombucha for the best substitute for alcohol and soda. Still, there is some sugar in water kefir, so don't abandon good old water as your main beverage choice!

Milk-based kefir

Milk-based kefir (usually referred to as simply kefir) is made from cow's, goat's, or sheep's milk. Milk kefir, like water kefir, is loaded with beneficial bacteria that exceed other foods, even yogurt, which is high in comparison to other fermented foods.

The sugar in milk, called lactose, is broken down during the fermentation process like the sugar is when making both kombucha and water kefir. So kefir is left with only about 1 percent milk sugar. Goat and sheep kefir contain even less lactose. Kefir also contains active lactase enzymes to break down even more lactose, so people with lactose intolerance are often able to digest kefir easily.

Nutrient dense

Valuable protein, vitamin D, and calcium come along with kefir (as well as yogurt). Milk-based kefir is also high in tryptophan, an amino acid that tends to be calming. Dairy products made using bacteria tend to negate issues found in regular dairy, so don't throw all dairy into the same category. They aren't the same to your body, and the health benefits of cultured or fermented dairy exceed those of regular dairy products.

Kefir made with milk isn't a replacement for water, but if you're going to drink your food, kefir is my choice. Plain kefir or lower-added-sugar kefir is a healthy addition to a smoothie or a smoothie bowl. There are so many microbes feeding on the milk sugar to make kefir that it ends up pretty sour. So beware of the added sugar used to cover that sourness. If you're replacing a soda or shake, a lower-sugar sweetened kefir drink is a much better choice for your health.

ALCOHOL

Alcohol temporarily keeps your body from burning calories while it is being metabolized.[24] Since your body is unavailable to burn the calories from the food you've eaten recently or you eat while drinking, the calories are stored as fat. Alcohol especially increases fat storage in the abdominal area, and thus is born the "beer-belly."

Over time, drinking alcohol can easily add fat pounds to your weight and all the health issues that go along with that. Consider ways to not drink when possible. There is a "sober curious" or "sober sometimes" movement happening around the country. It started as a social media challenge with New Year's weekend, which leads to "Dry January." Now there's "Dry July" and "Sober September" along with nonalcoholic social groups popping up all over the country. There are even bars opening in some cities called Sans Bar that provide a comfortable place where people can make sober friends, listen to music, and enjoy nonalcoholic drinks. This movement has spread across the country. People are challenging others to experience life without alcohol.

When you do choose to drink alcohol, here are some tips to help keep the pounds off.

Eat health-promoting foods

Eat high-protein, high-fiber foods before drinking so that the alcohol will be absorbed more slowly into the bloodstream and keep your blood sugars steady. Drinking alcohol boosts your appetite while reducing your willpower to eat healthier foods.[25] High-glycemic foods cause your blood sugars to drop, causing you to crave more of these foods. After a certain amount of alcohol, which is different for everyone, you're not going to care about having more fries or another drink.

Choose foods with protein, fiber, and healthy fats (like avocado, salmon, vegetables sautéed in olive oil, and peanuts) BEFORE you take that first sip!

Limit alcohol to one drink

Limit alcohol to one drink a day for women and two for men as recommended by the USDA Dietary Guidelines. These guidelines, based on the latest research and redone every five years, are very specific to recommend that nondrinkers shouldn't start drinking alcohol.[26] These guidelines also don't mean that you can skip drinking all week and save all your drinks for the weekend. That would be the worst scenario for your weight and your health.

Several drinks in one night have a much bigger effect than one drink a day. Three or four drinks in one night means that your body has several hundred calories from alcohol to metabolize before it can burn your food for energy. The more alcohol you take in, the longer it takes for your body to get back to metabolizing your food for energy. By that time, much of the food you've eaten will be stored as fat instead of burned for energy. You're also hungrier and less likely to be in control of your eating. The extra calories alone will eventually add on the pounds.[27]

ONE ALCOHOLIC DRINK EQUIVALENTS

One alcoholic drink equivalent contains 14 g (0.6 fl oz) of pure alcohol. Examples include:

- 12 fluid ounces of regular beer (5 percent alcohol)
- 5 fluid ounces of wine (12 percent alcohol)
- 1.5 fluid ounces of 80 proof distilled spirits (40 percent alcohol)

health.gov/dietaryguidelines/2015/guidelines/appendix-9/#table-a9-1

Choose alcohol wisely

Stay away from the fancy drinks that tend to have more calories, especially sugar calories that can make you even hungrier. Mix your drinks with some seltzer water and lime to dilute the calories. If you have to have a drink, have something simple like wine or beer. Vodka, gin, or bourbon with club soda are better choices than sugary cocktails. Club soda and seltzer water that don't contain calories or sugar dilute the alcohol as well as your desire for less healthful, high-calorie foods. Avoid sweet, syrupy liqueurs, juices, tonics, colas, and high-sugar bottled mixes for drinks like daiquiris and margaritas.

Drink water

Alcohol is dehydrating. It interferes with the regulation of water in your body. Each shot of alcohol you consume makes your kidneys generate an extra 120 milliliters of urine on top of the regular 60 to 80 milliliters an hour.

Unfortunately, you can't just drink more water to compensate for the extra water loss. You hang on to only about half or a third of the extra water you guzzle. You lose most of the water in your urine, and you still become dehydrated after a night of drinking. You will be better off than if you didn't drink any extra water, but you will still be dehydrated.

Hangover symptoms are mostly due to dehydration. Drink a big glass of water for every alcoholic drink you consume. If you've had quite a bit to drink, try to drink several glasses of water before going to sleep. Keep water next to your bed in case you wake up in the middle of the night. Drink water when you wake up. If plain water is making you nauseated, sports drinks may be easier to get down.

You might find yourself craving large amounts of greasy fast food. Being dehydrated can make you feel hungrier than normal. Also, your body needs the energy to help with the effects of drinking, so craving high-calorie greasy food is normal but just adds more unhealthy calories to the mix. Try satisfying those cravings with healthier fats like avocado toast and eggs.

CUT DOWN ON ALCOHOL

Over time, alcohol inevitably causes you to carry extra fat weight. More and more people are choosing to not drink or to cut down on their alcoholic drinks. Make seltzer with a splash of cranberry juice your standard order at bars, at least on either side of a drink order.

Take a walk

A movement like a brisk walk will help increase your oxygen level and will increase your metabolism to help clear out the alcohol and its by-products.

Bottom Line

Consider not drinking at all or at least drinking less. The damage caused by drinking too much is greater than the benefits of drinking a little, so don't drink multiple drinks in one day. Limit yourself to one alcoholic drink if you're a woman and not more than two drinks if you're male. Also, don't drink more than one drink per hour, which is the maximum speed at which your liver can process or detoxify alcohol. Start with a health-promoting meal and water before you drink any alcohol, and make a point to drink water in between alcoholic beverages. Keep your alcohol choices simple without the extra sugar, and decide which foods you will eat before you start drinking. Sparkling water can go a long way!

HYDRATE FOR FITNESS

Hydrating and fueling make a huge difference to your energy level during and after exercise. Without proper fuel and fluid, you're going to burn out much faster!

When you sweat, you don't just sweat water but also valuable electrolytes. These electrolytes (like sodium, potassium, magnesium, calcium, phosphates, and chlorine) have electrical charges that activate the nerves and muscles to function. Electrolytes keep your body systems working. When you are low in these critical minerals, the signals aren't sent correctly and cause cramping, headaches, confusion, dizziness, and poor performance among other potential problems.[28]

Sodium is the primary electrolyte you lose through sweat, but potassium and other minerals are lost too, just in smaller amounts. So if you're sweating a lot for extended periods of time, you must replace your fluids and electrolytes. The fluids lost can easily be replaced by drinking water, but to replace the electrolytes you must take in foods and drinks containing those minerals. Most foods and drinks in their whole-food form naturally come with those minerals in differing amounts, so just eating healthier foods or drinking milk will most likely do the trick.

FOODS HIGH IN ELECTROLYTES

Sodium – salted nuts and peanuts, pickles, peanut butter, soup, tomato juice or vegetable juice, canned tuna and salmon, deli meat, and cottage cheese

Potassium – white beans, potatoes, oranges, bananas, avocado, yogurt, tomatoes, spinach, red peppers, salmon, clams, and sweet potatoes

Calcium – yogurt, milk, cheese, cottage cheese, beans, lentils, and kale

Magnesium – seeds, leafy greens, nuts, canned salmon, legumes, avocados, bananas, whole grains, and dark chocolate

Chloride – celery, lettuce, olives, tomatoes, leafy greens, seaweed snacks, and table salt

Water, yogurt, and a banana can make all the difference. Maybe add some salty peanuts to the mix as well. Besides replacing your electrolytes, you're also replacing your energy stores with the carbohydrates from the yogurt and the banana. The yogurt and the peanuts will provide protein to help repair and build your muscles.

Sports drinks

The industry selling sports drinks wants you to believe that you always need these drinks when you exercise and that they are the healthy drink choice all the time. They have been pretty successful in their marketing campaigns since that seems to be a general belief.

You don't always need sugary sports drinks when you exercise. In fact, artificial colors and flavorings and sugar come along with the electrolytes. So if you're not sweating that much and exercising for an hour or less, skip the commercial sports drinks altogether and drink water. If you are sweating, drink water and eat something salty. Other alternatives for sports drinks like milk or drinkable yogurt work great too.

COCONUT WATER

Coconut water's high amount of potassium makes it popular for hydration during exercise. However, it doesn't have enough sodium to replace the sodium lost during heavy sweating. Orange juice and salty food like peanuts or nuts would actually be a better and less expensive choice.

Endurance sports drinks

When you're sweating excessively while exercising for an extended period of time, that is when you may want to opt for a sports drink. Dehydration can be dangerous! Make sure it's an endurance sports drink. Sports drinks don't generally have enough salt in them, so make sure the label specifies it's for endurance. Even then, you may need to pair it with a salty snack. The endurance sports drink at least replaces fluid lost along with the electrolytes and has an added bonus of additional carbs needed for quick energy if your energy stores are depleted. Try to buy one without added colors if you have a choice.

My team played in a tennis tournament several years ago in Palm Springs. The temperature on the court was 107°F. If we had salty foods or pickle juice to accompany our water in between games, that would be great. But we were out of town, and we didn't. It was a perfect time to drink an endurance sports drink. In fact, one member of our team drank water during our match instead of a sports drink. It was so hot, and we sweated so much that we all lost plenty of electrolytes in our sweat to warrant a sugary sports drink. As she drank more and more water during the match, she actually diluted the electrolytes in her blood, making her condition worse. The rest of us recovered in about fifteen minutes. Our water drinker took hours to feel good again. An important lesson for us all.

Bottom line

Listen to your body! If you exercise for sixty minutes and you're not a heavy sweater, you don't need anything more than water. Your need for electrolytes is directly related to how much you sweat. Even when you're exercising for ninety minutes and you're not sweating much, you're probably in more need of water than electrolytes. In fact, if you're not moving for hours at a time but you still sweat a lot, consume water and a salty snack as a precaution. Drink plenty of water before you start exercising, and drink water during your activity when you are thirsty. You don't need to overload on water if you feel good. If you're sweating excessively for more than an hour straight or moving in really hot, humid temperatures, eat or drink something with sodium to keep your electrolyte levels up.

SWEAT TEST

When you're doing light exercise in a dry, cool place, you'll probably only lose about 250 milliliters of fluid an hour, whereas if you exercise in a hot, humid place, you can lose 2 to 3 liters of fluid an hour. That's a big difference! Your need for electrolytes and fluid is directly related to how much you sweat.

FIX 7

WATER WELL

Drink water throughout the day and evening.

Train yourself to hydrate all day long. Keep water ready to drink within arm's reach. Leave a glass of water on your desk or your kitchen counter. Every time you notice it, take a drink. Water fills you up, and it doesn't add calories as food does.

Hydrate to burn

Water is necessary to burn body fat. Cells in your body need to be hydrated to do their job. Cells need to be swollen with water to burn body fat more efficiently. When our cells are low in water, they store more fat instead of using it for energy. Of course, once the cells are completely hydrated, more fat won't be burned by drinking more water. But drinking water consistently ensures that the fat cells can burn to their full capacity.

How much water

To check periodically how you're doing in terms of water intake, take a look at your urine. It should be clear to a very faint yellow color. Half of your weight in pounds is the number of ounces of water you need to drink throughout the day (e.g., 150 lbs. ÷ 2 = 75 oz. of water). But keep those cells hydrated all the time by drinking water throughout the day rather than at one sitting!

Beyond plain water

Decaffeinated tea, herbal tea, infused water, seltzer water, and flavored water are also good ways to replenish fluids, as long as they don't contain sugar or sugar substitutes.

If straight water isn't working for you, add a little fruit and mild vegetables like cucumbers to your water pitcher for some extra flavors. Herbs or fresh spices like mint or fresh ginger root add flavor too, but know that over time the flavor strengthens. If it's fizz you need, add some club soda to your water.

If you're feeling extra motivated, try adding a few drops of high-quality lemon or lime essential oil to your water. You might find yourself drinking twice as much. Even better, these oils may neutralize contaminants in your water and boost your defenses against illness.

Action plan

- ✓ Have water with you twenty-four hours a day, seven days a week.
- ✓ Leave a glass of water on your desk and kitchen counter. Every time you notice it, take a sip.
- ✓ Carry a water bottle with you wherever you go, and sip often.
- ✓ Drink water before you exercise, at relevant intervals during exercise routines, and after exercising.
- ✓ Avoid drinking caloric beverages like soda, energy drinks, juices, alcohol, and so forth.
- ✓ Eat fruit in whole fruit form instead of drinking fruit juice.
- ✓ Drink a big glass of water in the morning right when you wake up and right before you go to sleep.
- ✓ Add one or two drops of lemon or lime essential oil to your water.
- ✓ If you are transitioning to water from regular or diet soda, find a flavored water without sweeteners to help you make the switch.
- ✓ Infuse your water with fruit or herbs for more taste and nutrition.
- ✓ Make seltzer with a splash of cranberry juice your standard order at bars, at least on either side of your drink order.
- ✓ Track your progress on the *12 Fixes to Healthy* app.

FIX 8
SHIFT SURROUNDINGS

Weight and health success is built on a house of cards and ready to collapse with the tiniest disturbance to the popular diet plans. Instead of building your success on shakes, bars, supplements, menus, workouts, and difficult programs that are expensive and need to be followed for the rest of your life, skip the crash diets. Focus on your routines. Results will come more easily than you realize, without much thinking required!

Fix 8 is all about making small discreet changes to your environment that will help you to avoid temptations and to make the healthier choice into the easier choice. I call it "eating ergonomics." As a result, you will form new rewarding routines!

There are so many easy ways to shift your surroundings to form new routines. Let's split them into four categories—rearrange, resize, plan, and search.

CREATING NEW ROUTINES!

Habits are created by doing something repeatedly (obvious, I know!). Our brains make a strong link between the situation or environment where we do that behavior and the upcoming reward we get for doing it. A cue leads to an action that in turn leads to a reward, cementing a connection between that cue and action. It goes like this: turn on the TV before bed, feel a little hungry, eat some chips, and get a nice dopamine kick from the flavor and carbs. Tomorrow, turn on the TV, feel hungry, eat chips, get dopamine. After a time, it won't even matter that you are full after dinner; you will want to eat those chips when you turn on the TV.

So, take a behavior you want to change. Look at what the "cue" might be. A situation, a place, a certain friend, a time of day, or maybe a particular emotion. Be aware of what your cues are. All you need is enough awareness so that when a cue occurs, you can notice and shift your response.

You need to decide beforehand what that new action will be in response to that cue. Instead of TV leading to chips, maybe it will be TV leading to fruits or vegetables. To support this new choice, you can put that bag of chips in a spot that is difficult to access, and keep some fruit, veggies, and dip ready and easy to grab when you're feeling tired.

Rewarding routines

However, new routines need to be rewarding! You may not be happy going from eating chips to vegetables. But you may enjoy some strawberries or a cup of cinnamon tea. Shift your focus from trying to control your behaviors to setting up a new routine and reward instead.

Try different routines and rewards to see how you respond. Then become a mad scientist and test out different routines and rewards to see what helps the right behaviors stick. If one doesn't work, try another one of the many great suggestions sprinkled throughout this chapter!

REARRANGE

The right setup keeps you from draining your precious willpower unnecessarily. You make over two hundred unconscious food-related choices every day. You have to automate lots of your behavior because your brain can't possibly manage consciously making the thousands of decisions you make every day. Don't waste your willpower where you don't need to.

Your surroundings and routines drive so much of what you do and eat. They are the key to long-term success. The Cornell Food and Brand Lab demonstrated just how much our surroundings affect our eating. They gave people who had just eaten dinner at a restaurant some free movie tickets and buckets of five-day-old stale popcorn.

What happened? They ate it! Not because they were hungry or because it was particularly crave-worthy. Even more, those who got large buckets ate 34 percent more popcorn than those who got smaller buckets. They, like all of us, ate because of the environment around them.[1]

Intentionally putting things in certain places can make the wrong decisions less convenient and the right ones more convenient. It can remove temptation and feelings of deprivation while making the healthy choice more instinctive and, eventually, habitual. The strategies are simple. Yet they make a big difference! Here are some ways that you can arrange your environment.

In plain sight

Researches found we are three times more likely to eat the first thing we see compared to the fifth thing we see.[2] That means that your fridge, cupboards, counters, and workspaces need to have ready-to-eat, healthy options, screaming for your attention. Put the right foods in plain sight by leaving fruits and vegetables on the counter, on the kitchen table, on the front and eye-level shelves of your fridge, and in see-through containers. Keep those cupboards full of nuts and whole-grain snacks. Keep health-promoting shelf-stable snack options in your car or on your desk.

Out of sight, out of mind

The more you see food, the more it is queuing in your brain, begging to be eaten. One study reported that women who kept even one box of cereal on the counter were twenty pounds heavier than their neighbors who kept all cereal out of sight. If it was soda they kept on the counter, they were twenty-five pounds heavier. But if it was fruit they kept on their counter, they were thirteen pounds lighter than their neighbors who didn't.[3] In another study, administrative assistants with a bowl of candy on their desks ate an average of nine pieces per day and weighed 15.4 pounds more than those who kept the candy bowl six feet away and ate only four pieces a day.[4]

If the off-limit food stays at the grocery store, all the better. If it manages to make its way home, store it in an annoying place, such as a garage or storage room. I moved my chocolate chips from the pantry to the downstairs storage room when I realized that I was grabbing unnecessary handfuls more often than I would if I didn't see them sitting there on the shelf. Place food that we should eat less of in out-of-sight, inconvenient places and in opaque containers. This keeps us from feeling deprived as often because we're taking away the temptation before it begins!

Immovable surroundings

Sometimes you may need to rearrange yourself in relation to possible choices. For instance, if you tend to stop for a burger, fries, and soda on the way home from work, you may want to change your route home, even if it takes a little extra time. You obviously can't change where the food establishment is located, but you can reroute yourself. You can form a new routine with healthier choices that are more rewarding!

Serve up some magic

It always feels impossible to get yourself and your family to eat more vegetables. Some tweaks to your surroundings can seem like magic happening in front of your eyes. Here are three ways to make vegetables disappear.

Pre-meal vegetable platter

Set out a plate of cut-up vegetables, just in case anyone is hungry. This is especially effective before mealtime when you and others are the hungriest. Those vegetables will disappear as if by magic.

When kids (both big and small) complain that they're hungry before a meal, don't let them make a beeline for the pantry. Give them two choices. Tell them that they can eat any of the vegetables on the table or nothing. But it is their choice, so it's not a power struggle. They will choose to eat their vegetables. Be consistent and they will stop complaining. If you want them to try something new, put it on the vegetable plate, and don't even tell them to eat it. People are a lot more likely to try new things when they're at their hungriest and when they don't feel like they're being forced to eat them.

You don't have to put a lot of extra time and effort into cutting things up. You can buy vegetables that are already cut or don't need much cutting. Sugar snap peas, baby carrots, and grape tomatoes pour right onto the plate. Cucumbers and red peppers are quick and easy to cut. Whatever you choose, I recommend having more than one type of vegetable out there so that people have more choices and more exposure. Hummus or salad dressing may be nice encouragement at first, but as time goes on and eating vegetables becomes a habit, you can skip the dressing altogether, and the veggies will still get eaten. Routinely add an unfamiliar vegetable to the mix without mentioning it. You may be surprised about what gets eaten.

Serve vegetables first at mealtimes

This habit can make a huge difference for your family's vegetable intake. Whatever vegetable you planned for dinner, make and serve it first at mealtimes. People will eat more of it. Why? Because they're hungrier at the start of the meal. Don't put anything on the table to compete with your vegetables. The salad or the broccoli goes on there first and alone. Add the other food only after the vegetable is gone or everyone has had plenty.

Once you've served the green salad or vegetables at dinner, serve portioned food on the plates rather than serving "family style." Researchers have found that this strategy helps us to eat 19 percent less food. If you need to use serving bowls, be sure to take them off the table after everyone's first helping—except for the fruits and vegetables! When food is within reach, we eat more of it. A family-style dinner, where everything is in reach, promotes overeating.

The produce-only rule

We eat more when we're doing other things. Use distracted eating as a way to eat more vegetables. Serve only fruits and vegetables for TV time, while doing homework, or while playing games. Spend one or two minutes chopping up a vegetable or fruit for your family that they may not typically snack on. Do this for yourself too! Put the cut produce next to yourself while you're browsing the web or next to a family member while they're busy doing their thing. That produce will disappear. Magic! Use fruits and vegetables to replace crackers and chips. As long as there's no processed food competition, the fruits and veggies will disappear.

When I buy pears and leave them washed on the counter, no one eats them. But if I take ten seconds to cut a pear with an apple cutter and set the slices next to my kids, then voilà: the contents of the plate magically disappear. We will devour the fruits and vegetables without realizing it, just because they're there. Not only that, but after a while, everyone starts to get used to them. We all realize that those pears are actually delicious. And once in a while, who knows? You might even see a family member grabbing one from the fruit bowl or refrigerator. My young adult kids do almost every day now.

Consistent guidelines or rules help too. Make a family rule that only fruits and vegetables can be eaten in common areas such as the TV or game room and all other foods have to be eaten in the kitchen, sitting at the table. This promotes a more mindful approach to high-calorie snacks, which helps you eat less of them. It's more difficult to overeat produce than to overeat a bag of chips. This is a great strategy for supporting a healthy weight for yourself and your kids. Not only are you eating more healthy food, but you're also eating less unhealthy food at the same time. With a rule like this in place, people will eat fruits and vegetables without thinking twice.

Your surroundings do affect what and how often you eat. Use this technique to make eating healthier easier, not harder. If you eat lots of fruits and vegetables, you're less likely to eat or even want unhealthy food.

Investigate your surroundings

Walk into your kitchen or look around your office or cubicle. What are the first and most visible options? Are they nutritious? Are there some items that need to get pushed to the back of your shelves or drawers for less frequent consumption?

WHAT COLOR IS YOUR PLATE?

Even the color of your plate can affect how much you eat! Plates with a similar color to the food served may cause you to eat 18 percent more food than opposite-colored plates. What color are your plates? Choose foods that are not the same color as your plate—unless you're eating a vegetable.

RESIZE

How and when you serve your food can make eating healthy effortless. It's amazing how you can be more satisfied with less food with simple changes to the size of your dish, package, or portion. These changes are so easy that you may not believe these techniques actually work!

Dish size

Remember the team from Cornell? They found that smaller plates lead us to think we're eating more food than we really are. In fact, participants of the study filled their plates about 70 percent full—regardless of plate size.[5] So if you're using a ten-inch plate rather than an eight-inch plate, you could have about 50 percent more food on your plate. More food on your plate translates into more food that you're likely to eat.

In other studies, people served themselves more ice cream using a larger spoon[6] and ate less food when using a smaller fork.[7] Glassware has the same effect. Tall, narrow glasses make us feel like we're drinking more.[8]

Harness this powerful illusion. Trick your brain into thinking that you're eating more, and use smaller plates, bowls, silverware, and glasses. The same portion will look bigger with a smaller dish, so you're likely to eat less, with more satisfaction.

Make the switch

So what does this tell you? Use small plates for food that you need to eat less of. Use larger plates to eat more of what you should. In fact, get in the habit of switching plates; use the larger dinner-size plate for your salad and vegetables and the smaller salad or dessert plate for the rest of your meal. Apply this same principle to your other dishes. Use smaller bowls and utensils to eat everything else but your produce. Use shorter, wider glasses for water so you hydrate more. Put less healthy and sweetened drinks in tall glasses so you take in less.

JUNIOR-SIZE PORTION

If you do end up in an ice cream shop or a fast-food joint, buy a junior cone or junior or kids' size hamburger. Enjoy each morsel while you satisfy your craving with fewer calories. I do this in ice cream stores all the time. I don't feel deprived at all, and I do savor every bite!

Portion size

Think junior-size portion all the time, especially when you dish up your food yourself and don't have a choice in the dish size. Remember, your eyes are bigger than your stomach. Use your brain instead of your eyes, and consciously dish up less food, unless it's produce. You can always go back for more if you still want it. Let your body's fullness level dictate your decision. If it's on your plate already, you are more likely to eat it! You want to eat more produce and drink more water, so reverse this technique for those items.

Package size matters

Eating out of large packages leads to eating more than if you have a smaller package, so dish your food out of the package into a small cup. Better yet, repackage food into single-serving portions. Bigger ingredient packages may encourage you to cook 22 percent more food.[9] It's okay to put half of the food away in the refrigerator or freezer for another meal before you serve up.

SNACKING RULE OF THUMB

Make a plan to put your snack in a smaller container, then put the original package away before you mindfully indulge!

If it's a food that comes in small pieces like nuts or popcorn, put just one nut or popcorn kernel in your mouth at a time. Chew that completely before you eat the next bite. If using a utensil, take the smallest bite possible. By not eating from the original package and chewing small bites thoroughly, you will enjoy your food more and end up eating less!

PLAN YOUR SURROUNDINGS

You bought fruit and left it on your counter. But then nobody ate any, and they went bad!

Just having fruits and vegetables on the counter isn't the only step to success. Make a plan to cut up your fruits and vegetables until it becomes part of your routine. Buying healthy foods won't do you any good unless you actually eat them. Foods that fuel your leaner body and stronger mind might be sitting a few feet away from you every time you walk into your kitchen. But unless you come up with ways to close the gap between the counter and your stomach, your good intentions become useless.

That's why it's important to build some practical tricks and habits for transforming distant vegetables into food that you actually eat. Don't despair: it is easier than you might think to have nutritious meals ready with minimal effort. Plan out the meals for the week in advance, make extra food, and prepare for those unexpected chaotic days that leave you either tired or short on time.

Meal plan

The simplest place to start is by planning a one-week menu, with recipes and a grocery list. If you don't typically plan meals, maybe start with planning one meal a day for a week and build from there. When trying to incorporate new and healthier foods, be sure to include some familiar recipes you know you enjoy. You may have had times when you were trying really hard to be "good" and planned a week of new, super healthy-sounding recipes . . . only to find that you really didn't enjoy them. Halfway through the week, you and your family felt like going out to eat because you wanted something comfortable.

Tasty health-promoting recipes

Look for recipes that make a few changes at a time to the types of food you are already used to. Nutrition needs to be enjoyable to be sustainable. Make a point to make your meals flavorful and satisfying, not just healthy. Check out my recipes at foodswithjudes.com or other sites to find new, appealing recipes that are also health-promoting. There is a meal planning app for healthy eaters called Mealime that my clients swear by. It has lots of good recipes too.

Batch prep

Use weekends or lower-key days to make a large batch of delicious healthy food so you have prepared leftovers in the refrigerator for lunches and frenzied evenings. Use these days each week to make a large pot of soup, wash lettuce, and cut up vegetables for meals and snacks. Planning is an investment of some time that can pay off later. Get the most out of this time and energy by doing bulk prep and cooking. Cook double portions for at least one to two meals per week. Either freeze or refrigerate in single portions, ready for later.

When making rice, quinoa, or pasta for dinner, cook extra and put it in the fridge or freezer to make grain bowls, stir-fries, and hot cereal. Or you can quickly pair precooked whole grains with soups, lentils, and vegetables for a quick weekday meal. Other items that store well and can be cooked in bulk include soups, hearty salads, stews, bean chili, enchiladas, breakfast burritos, egg muffins, muffins, fruit sauce, hard-boiled eggs, and fruit and nut energy bites.

> **BATCH COOKING TIP FROM DIETITIAN ALLISON**
>
> When you make a batch of soup, chili, or stew, freeze leftovers in a muffin tin. When frozen, pop them out of the tin and put in freezer bags (make sure to get out as much air as you can). When you're in a hurry, thaw one or two of the frozen soup pieces and heat for a quick addition to your meal.
>
> Allison Harrell, RDN, CCMS, Virtual Dietitian at www.mywholenewlife.com

Real-life plan

Life does not go as planned, and extra time can be hard to come by. This affects what we eat. Think about your current go-to meals and snacks. Everyone has them. When you're tired, rushed, and hungry, what do you eat? Do you go out for lunch? Do you rely on the vending machine to get you through the day at work? Is there a thirty-two-pack of snack-sized potato chips sitting front and center in your pantry? Do you resort to fast food or quick alternatives such as boxed macaroni and cheese or pizza delivery for dinner?

Go-to meals

Have three healthy go-to meals as backups off the top of your head. These meal ideas can just be healthier versions of your current regular default meals. If quesadillas are your backup meal now, change the flour tortillas to a soft corn tortilla and add black beans, a little sharp cheddar, great salsa, and avocado, for instance.

New or normal, it's crucial that you choose your go-to meals in advance and be prepared when you need them. Store the nonperishable ingredients on hand. Keep the perishable items on your mind (better yet, in your smartphone) so you know exactly what items to purchase when you are scurrying into the grocery store, tired and rushed. Remember, three is the magic number for go-to meals.

Go-to snacks

Go-to snacks are also essential to keeping up good nutrition when life throws you a curveball. Keep at least three go-to snacks on hand for just these moments. And always keep one in your bag for the unexpected situations that arise.

When you're on the go and in a hurry, it's hard to resist grabbing prepackaged foods on the way out the door or while you are out. That's why having a few easy snack ideas already on hand or with you is essential to getting you through those hectic times.

Improve your health

Your snacks are a great opportunity to bring more fruits, vegetables, and protein into your diet. If you snack on real food in its whole form, it can do wonders for your health. On the other hand, if your snacks are processed carbohydrates, they will be broken down and digested quickly, leaving you with a big rush of insulin, making you hungry again sooner, and putting your body in fat-making mode. To add insult to injury, you are more likely to crave more processed junk.

Even when you're short on time, get in the habit of eating healthy snacks! Arm yourself with great-tasting, healthy snacks at work, on the road, and at home to avoid unhealthy temptations. This is one healthy habit in particular that makes a big difference to your body over time.

The snack box: simple snack ideas

Frozen fruits make easy and delicious snacks. Mangoes from the frozen section are one of my favorites. Eat them straight out of the bag—like a healthy version of packaged candy. In fact, they are my nineteen-year-old daughter's favorite sweet snack. A couple of years ago, she got her whole tennis team eating them, and it was easy for me to buy them, throw them into a cooler bag, and bring them to matches when it was our turn to bring the snack. You can try surprising everyone at work with mangoes instead of donuts. The tennis girls certainly liked the mangoes over donuts.

I love to nibble on pomegranate seeds and add them to my yogurt and hummus. You will never taste hummus the same way again. Use cucumbers to eat the hummus instead of pita chips. Fresh pomegranate seeds come out so easily when you hit the pomegranate with a wooden spoon.

People also swear by frozen grapes. When my bananas start to turn brown, I peel them, chop them into slices, and freeze them. Then I eat them plain or with a dab of natural peanut butter.

LIMED APPLES

Have you ever tasted a limed apple? My kids never used to like apples in their lunch. They wanted them cut, but they wouldn't touch them once the slices started to turn brown. After I started dipping the apple slices in fresh lime juice, they miraculously started to love them. The apples don't turn brown, and the lime juice lasts a few days in the fridge, so you don't need to squeeze new lime juice every time. While I was at it, I always made some for myself too. Yum!

Toss some frozen bananas into a food processor with some strawberries, almond butter, or cocoa. Let it blend until it's the consistency of soft-serve. It will satisfy your ice cream cravings—the healthy way!

Dates are another great option. Eat them plain, or stuff them with cheese, nut butter, or nuts to make a filling snack or a decadent appetizer. Fresh figs are amazing too.

Enjoy hummus with cucumbers, red peppers, or celery. How about eating edamame or yogurt? Pistachio nuts in their shells force you to eat them slowly as another healthy option. Whole-grain crackers (with just a few ingredients), like Wasa Light Rye crackers or the Hint of Salt Triscuits, topped with a thin slice of artisan cheese or cottage cheese and cucumbers are on my go-to list as well.

50/50 SNACK TIP

Use my 50/50 snack tip and make your between-meal bites half protein and half produce. This snack method helps you eat enough protein throughout the day to build muscle, but it does more than that. The produce will provide carbohydrates to be used for energy, so all the protein you eat can be saved to preserve and build muscle. All the amazing nutrients in fruits and vegetables will also help your body in so many ways, now and in the long run. Remember, eating fruits and vegetables is your best daily "detox cleanse."

Workday wins

When it comes to eating healthy, the office has a way of sabotaging our best intentions. Someone brings in donuts. You work through lunch to make a deadline. The vending machine stares us in the face from where we're stationed in our desk chair or on the way to the bathroom. Victory over these temptations is possible, with a little planning.

There are plenty of convenient whole-food snacks to take to the office. The key is picking healthy foods that you really enjoy and then making sure you have them on hand.

Make your snack plan

Decide today which snacks you are going to try (see exercise above). Then stock some snacks in your office. Even better: put some in plain sight on your desk. As soon as you notice hunger, fuel up so that extreme hunger doesn't weaken your willpower or decision-making. An apple sitting on your desk is a good example. No one is overindulging on apples. But paired with a handful of nuts you keep in your bag, an apple can make all the difference by preventing your desire to indulge on the donuts.

> **APPLE OF YOUR EYE**
>
> Keep an apple on your desk at work. If you're hungry, you're more likely to eat it just because you can see it.
>
> Have you ever tried peanut butter with your apple? Keep an apple cutter in your desk drawer at work and maybe even a container of peanut butter. It's very satisfying! Just give it a try, and see what you think!

Over time, not only will you get used to bringing and eating produce at work but you'll enjoy them even more than those donuts or anything you could buy from that vending machine. You will recognize how much better you feel when you eat fruits and vegetables—and how poorly you feel by comparison when you resort to junk. Your tastes acclimate to less salt and sugar as well, so you're not as enticed, but the real trick is getting into the habit of keeping healthy food with you!

Self-talk

When everyone is grabbing donuts that the boss brought in and you decide not to join in, don't tell yourself how much you are missing out. Flip the conversation in your head from "I'm missing out" to "they are missing out on feeling awesome." You can still enjoy the festivities and conversations without eating junk. This is especially true if you stay filled up on nutritious and yummy meals or snacks.

Pre-dinner snack

I have several corporate clients who make sure they don't go to business dinners hungry. They will eat a handful of raw almonds and a piece of fruit, for example, before they go. It seems counterintuitive, but if they don't eat some whole food before they go, they are less willing to listen to their own bodies' satiety cues throughout the evening.

Dinner ingredients delivered

Having ingredients delivered to your door is magical. That's coming from someone who loves to cook. These services certainly make cooking easier and more fun for everyone, but especially for the novice. These ingredient-delivery services are popping up right and left and are another terrific option for your busy life.

There is a cost for the convenience, but most meal kits cost less than what you'd pay in a restaurant for a similar meal. The cost might even be comparable to a meal at home if you are not good at using leftover ingredients or if you tend to buy fresh ingredients and let them go bad before using them. Expect to pay around $10-12 per meal.

These companies are everywhere now. The question is, which services are delicious and healthy? Look for companies that emphasize health and nutrition and include more Mediterranean-based meals!

SEARCH YOUR SURROUNDINGS

When eating out, healthy options are rarely front-and-center. You may think that if someone else is preparing the food, you're out of luck. You may tell yourself it is too difficult to eat healthy, so you don't even consider the options. The truth is that it's easier than you think to maneuver restaurants in healthy ways. Opportunities for good nutrition are there. If you look, you will find them.

> **CALORIES ON MENUS**
>
> FYI, chain restaurants have new menu-labeling guidelines that were initially signed into law in 2010. These new laws took effect in 2018. Calorie amounts must be displayed for all foods. Just make sure that your calories are filled with healthy whole foods that have not been processed.

Navigating restaurants

One corporate client of mine is a champion healthy restaurant eater. Not only does he know how to order but he has also figured out how to re-engineer his environment to help him make good choices. Here are ten tips that worked for him:

1. Use your water glass to block bread, appetizers, and desserts from your line of vision.
2. Only drink water until your meal comes.
3. Keep the bread or chip basket off the table or at least out of reach.
4. When possible, skip the bread, buns, and tortillas and ask for double vegetables instead of rice, potatoes, pasta, or french fries.
5. Use vinaigrette over creamy dressings and skip the croutons.
6. Order the salad dressings on the side and use it sparingly. Try dipping a forkful of salad into the dressing instead of pouring it over the salad.
7. Stop eating before you feel uncomfortable. Take the remaining food home for a future meal or snack when you are hungry and will enjoy it more.
8. Try to stick to fresh fruit for dessert, and when you have something else, order it for the table and enjoy a couple of bites.
9. Limit your alcohol intake and stick to water or seltzer and lime.
10. Order grilled seafood often.

HEALTHIER FOOD SHIFTS

CHOOSE...	INSTEAD OF...
Brown rice or an ancient grain like quinoa	White rice
Double vegetables	Potatoes or fries
Corn tortillas or lettuce wraps	Bread and flour tortillas
Fish, beans, lentils, yogurt, or nuts	Red meat

Don't be shy about asking for things to be prepared in healthy ways. Chefs don't mind if you ask them to skip the butter or cream in your order. Fish is a great example. Ask the restaurant to grill it with olive oil instead of frying it or cooking it with butter or cream.

Going casual

Choose fast-casual restaurants over drive-through chains. Fast-casual restaurants, from Indian to Mediterranean to Asian to Latin American fusion, are popping up everywhere. With apps and online ordering, it can really be just as fast! Plus, they offer us the freedom to choose what goes into our meals with fresh, quality ingredients. Don't miss the healthy obsession with grain bowls either. They are a perfect way to eat ancient whole grains with a lot of veggies in delicious, filling ways.

If you're at a Mexican grill, order a burrito bowl and choose a little brown rice over white rice. Add extra black beans, fajita-style fresh vegetables, salsa and pico de gallo, triple the lettuce, and occasionally some grilled chicken or fish. Not only is it healthy but I don't usually eat the whole bowl. I save the rest of it for another quick meal later when I get hungry again.

Even Chick-fil-A offers some quick and healthy options such as grilled chicken nuggets, a fruit cup, and a delicious kale salad. Chick-fil-A also serves oatmeal for breakfast, as do Starbucks and most other coffee places. No matter where you go, be sure to skip the white bread, buns, and rolls altogether. Wheat bread offered in eating establishments is often highly processed as well but is better than white bread. When eating out, choose vegetables over bread when possible!

PARTY TIME ATTACK PLAN!

Identify all your food options before you decide what you're going to eat at a party. When you go to an event serving buffet-style food, look over the choices before you get in line. Knowing all the choices being served helps you narrow down what to eat as you go through the line. Use the smallest plate or bowl available and take small portions. Remember that you can go back if you are still feeling hungry. Be choosy and fill most of your plate with vegetables first.

Food on the fly

As I'm running out the door on the way to my next adventure, I habitually throw in my standard travel items: a water bottle, apples, and nuts. I recently added pouched salmon in olive oil along with Hint of Salt Triscuits to my arsenal. There's no prep necessary, and I can take them anywhere without a hassle. This travel kit fuels me on a moment's notice and saves the day often. Keep your eyes open for whole-food items you can throw in your travel bag to make your trip better.

I fly a lot, and I am happy to see an increasing number of healthy options available at airports as well as trendy eateries offering more nutritious foods. One time I was stuck at the airport for six hours. But I stayed satisfied with a vegetable omelet for my breakfast and a salad with lots of vegetables, ancient grains, and protein for lunch. There are areas with lots of healthy options including a salad/food bar in the Newark Liberty and LaGuardia airports.

Most airports will have ready-to-buy packages of fresh vegetables such as celery and carrots. Eat these raw veggies with hummus or nut butter that is also available. Even hard-boiled eggs are being sold at many airports. You can almost always find Greek yogurt, cups of cut fresh fruit, and nuts in these grab-and-go stores, along with salads. Uncut fresh fruit (such as apples) as well as Kind Bars are often found near the register.

If you get on the plane in a starved state, there are often one or two healthier options. Delta has a fruit, nut, and cheese plate that is full of whole foods other than the crackers. Just try to stick with whole food, and skip the packaged food such as crackers, pita chips, potato chips, cookies, pretzels, and bread. Peanuts or almonds, even in their dry-roasted, salted form, are the healthiest snack options along with fresh fruit and vegetables.

Bon, voyage!

ON THE GO SNACK TIP

When selecting fruits and vegetables to go, choose durable produce (for example, an apple or orange) or pack more-delicate produce in sturdy containers. I can't tell you how many times having these durable apples on my travels has saved the day for me and whoever I'm with. It's not just in the airport or on the plane either. I'm often traveling in new places with unknown schedules and little access to healthy food. I whip out an apple from my purse, and it keeps me going until the scheduled meal.

Corner convenience

Believe it or not, you can actually find healthy options at your corner gas station. These stores are trying harder to sell healthier food and beverages. I recently attended a convenience store conference and was impressed with how store owners are trying to fill their shelves with sparkling water, flavored waters, probiotic drinks, and 100 percent fruit smoothies.

Convenience stores have more whole food choices such as nuts, cheese, Greek yogurt, salmon jerky, fruits, and vegetables. Look for a snack mix with edamame, peanuts, nuts, pumpkin seeds, and/or cranberries. With the exception of dark chocolate, skip the nut mixes with M&Ms or other candy thrown in. Some Kind Bars (the ones made with nuts and whole grains) are the better granola bar to buy among so many choices. Another option is a fruit bar made by That's It. They are fruit bars made with only two types of fruits and no added sugar. They even sell the bars with shelf-stable probiotics. If you need a little chocolate, they also sell a dark chocolate truffle filled with just fruit and no added sugar.

Even ancient grains are showing their face in this arena. Look for a roasted ancient grain snack called Kracklin' Kamut by Big Sandy Organics that is made from the ancient Khorasan wheat called Kamut. The Kamut is roasted, not fried, with safflower oil, which is high in monounsaturated fat to help burn fat, and sea salt. It's crunchy and tasty with 6 grams of protein, 3.4 grams of fiber, and no added sugar per 1.5-oz. serving.

Larger convenience stores even have cut fruit, turkey sandwiches made on wheat bread, and fresh vegetable salads. Keep your eyes open and remember which corner stores offer the healthiest fare. It's going to get better and better given that millennials tend to snack more often.

Whether you're in a grocery store or a restaurant, healthy food is everywhere. Look and you will find it! You just need to think outside the box.

BUDGET CUT

Many people think that whole food is more expensive than processed food. In fact, the opposite is true. Real foods in their natural forms are often cheaper than processed foods dollar for dollar. For example, a medium bag of potato chips is more expensive than most fruits and vegetables.

When foods are processed, they usually cost more. Especially when you consider what nutrition you are getting from processed foods. Foods in their natural form contribute far more nutrients for your buck than processed foods do.

Oatmeal

A great example is oatmeal. Old-fashioned oatmeal is very inexpensive when bought in a large bag. Half a cup of rolled oats costs less than half of the cost of an instant packet of oatmeal. The only difference is that it takes three minutes to cook in the microwave instead of one minute. It provides far more nutrition than a box of Frosted Flakes cereal does.

You have full control over the sweetener you add, and there are no artificial flavors or colors. If you are used to sugar-filled breakfast food, you may need more sweetener in your oats at the start. Over time, gradually reduce the amount of sugar and you'll find that you still enjoy it with less.

Packaged food

Carrots and sugar snap peas are cheaper than lunch-sized packages of cookies and crackers. Apples and peanut butter cost less than a pint of ice cream.

The same thing is true of rice. Half a cup of good brown rice (such as brown basmati or jasmine rice) is much cheaper than a prepackaged rice mix such as Rice-A-Roni or Near East. The difference is a few spices and maybe some bouillon. Do you really want to pay double for that? Take a few seconds to add your favorite spice mix. If you don't have one, buy a couple of mix combinations the next time you're at the store.

Beans and lentils

Beans and lentils (also called legumes or pulses) are inexpensive foods with supercharged health benefits. Pulses are the most cost-effective proteins and are staple foods throughout the world. The cost per serving of dried peas, beans, and lentils in the United States is 10¢, while beef costs $1.49, pork costs 73¢, and chicken costs 63¢. There are so many options to keep your meals interesting. You can find more exotic lentils, such as red, orange, or yellow varieties, at an Indian or Middle Eastern grocery store for much less than at a standard American store (if the latter sells them at all).

Popcorn

Popcorn is another great way to save. Microwave popcorn seems to be America's go-to snack. You are paying tenfold for someone to put the kernels in a bag and add salt and butter. Plain popcorn kernels are very inexpensive. It is also easier to find organic popcorn kernels as the demand for non-GMO corn grows.

Put a handful of popcorn kernels in an air popper or a paper lunch bag, fold down the top edge, and microwave it until you hear three to five seconds in between pops.

A hot air popcorn popper works too. Trade out the butter by spraying olive oil infused with lime or other flavors onto the popped kernels. Get creative with spices such as garlic powder, smoked paprika, turmeric, or chili powder. Or for a sweeter option, drizzle cocoa powder and either pure maple syrup or raw honey over popped popcorn. You won't just be helping your wallet; you'll be helping your health as well.

FIX 8

SHIFT SURROUNDINGS

Rearrange, resize, plan, search your environment.

Most of us eat without thinking. If it's there, we eat it. The more food that happens to be near us, the more we'll eat. We don't eat food because we're hungry. We eat it because it's available. And that has disastrous consequences for our weight and our health.

Rather than fighting your willpower to counter excess pounds, you can change your surroundings. Making small, seemingly insignificant changes to your environment will help you avoid temptation and make the healthier choice without thinking about it. Trick your brain to avoid the decisions that add on the pounds.

These little adjustments add up to greater success in your battle to eat and feel better. Let's look at four specific areas in your life to make environmental shifts—rearrange, plan, search, and serve.

Rearrange

Rearrange your surroundings to make healthier food front and center while keeping less healthy food out of the picture. The research reported that women who kept even one box of cereal on the counter were 25 pounds heavier than their neighbors who kept all cereal out of sight. In another study, administrative assistants with a bowl of candy on their desks ate an average of nine pieces per day and weighed 15.4 pounds more than those who kept the candy bowl six feet away and ate only four pieces a day.

Evaluate your surroundings. When and where do you tend to eat less healthy foods? Replace that food with cut-up produce, or shift your environment to avoid the sight of less healthy food.

Set a plate of cut-up vegetables near people while they're doing other things, just in case they get hungry. This is especially important before a meal. It's a great way for everyone to eat more produce.

Serve the vegetables first at mealtimes. You and your family will eat more of them since everyone is the hungriest at the start of the meal. If eating family-style, put the other food items on next but remove them after everyone is served. When food is in reach, we eat more of it, so keep only the produce on the table.

Use mindless eating to your advantage. You eat more when you're doing other things, so eat vegetables and drink water if you are on the computer or reading. Avoid eating unhealthy or neutral foods during activities.

Resize

How you serve your food makes a difference in how much you naturally eat. You are often more satisfied with less food when you make simple changes to how you serve your food. These changes are so easy that you may be surprised these techniques actually work!

Subconsciously, smaller plates and bowls lead you to think you're eating more food than you really are, so you are often satisfied with less food. Tall, narrow glasses make us feel like we're drinking more. Even plates with a similar color to the food served may cause you to eat more food than opposite-colored plates. Use a larger plate for the starter salad and a smaller plate for desserts. Use shorter, wider glasses for water so you hydrate more. Drink less healthy, sweetened drinks from tall glasses so you take in less.

You eat more when you eat from a larger package. So if you want to eat less with more satisfaction, eat from a mini package or dish the food into a small dish or container. Try repackaging foods into single-serving portions from the get-go.

It's an illusion, but it works. Trick your brain into thinking that you're eating more by starting with smaller portions with smaller dishes, utensils, and tall, narrow glasses for anything other than produce and water. The same portion will look bigger with a smaller dish, and you're likely to eat less with more satisfaction. Even if sometimes you go back for seconds, overall this method will help you to eat less food without fighting your willpower. You will have more time to decide whether you are full, and you'll need to get up and serve yourself more instead of having tons of food right in front of you. You'll probably grab some vegetables instead if they're sitting within reach!

Plan ahead

Plan ahead so you always have delicious, satisfying food to eat. The simplest place to start is with a one-week menu with recipes and a grocery list. If you don't typically plan meals, maybe start with planning one meal a day for a week and build from there.

Use weekends or lower-key days to make a large batch of a delicious healthful recipe so there are healthy leftovers in the refrigerator for lunches and frenzied evenings. Use these days each week to make a large pot of soup, wash lettuce, and cut up vegetables for meals and snacks. When making rice, quinoa, or pasta for dinner, cook extra and put it in the fridge or freezer to make grain bowls, stir-fries, and hot cereal.

Life does not go as planned, and time can be hard to come by. Think about your current go-to snacks and meals. Have three healthy go-to meals and snacks ideas along with the ingredients ready to go for when you're tired or rushed.

Search around

Look carefully at menus at nice and casual restaurants, and search for healthier food choices at airports and convenience stores. Instead of white rice, choose brown rice or an ancient grain like quinoa or more vegetables. Instead of potatoes or fries, order double vegetables.

Skip bread, rolls, and buns in favor of salads without croutons and dressing on the side. Go easy on the dressing. Instead of bread and flour tortillas, choose corn tortillas or lettuce wraps. Instead of red meat, choose grilled or sauteed fish or beans, lentils, yogurt, or nuts. Skip overly processed food, and choose foods in their whole-food form.

Action plan

- ✓ Eat without distractions such as computers, TV, or reading. If you do eat something while multitasking, go for vegetables (which you can cut up in advance for just these moments) and water.
- ✓ Start with a small amount of food. Use small plates, bowls, and utensils.
- ✓ Drink everything that isn't water from tall, narrow glasses.
- ✓ Keep unhealthy and neutral foods out of sight or in an inconvenient place.
- ✓ Place healthy foods in plain sight.
- ✓ Eat packaged foods in small servings. If you buy large packages, divide them into smaller packages in advance.
- ✓ Keep serving dishes (except for salad and vegetables) off the dining table after everyone has been served.
- ✓ Eat only part of a snack, and then wait ten to fifteen minutes. This smaller amount may be enough to satisfy you.

- ✓ Keep unhealthy food out of sight and out of reach in the office, and keep fruits, vegetables, and water in sight and within reach.
- ✓ Keep an apple on your desk at work and bring a few in your bag when you travel.
- ✓ Designate a time each week to plan out your meals, make a shopping list, and schedule time to prepare. Integrate at least one extra-batch recipe per week in your plan to prepare during less busy times.
- ✓ Cook double portions for at least one to two meals per week. Either freeze or refrigerate in single portions, ready for later.
- ✓ Avoid family-style eating with too many choices (unless it's just the vegetables and salad).
- ✓ Limit all eating to the kitchen or dining room. Always sit down at the table while eating, and never do other activities (except for talking, of course) while eating. This helps curb mindless munching in front of the screen, especially in the evening.
- ✓ Avoid buffet-style restaurants. But if you end up at a party or an event with a buffet, look at all of your food options first, before you dish up. Fill your plate with vegetables, and choose just a couple of small portions of other items.
- ✓ Take a water bottle and a few durable whole foods with you on all your travels.
- ✓ Track your progress on the *12 Fixes to Healthy* app.

FIX 9
SWITCH FAT

Some fats provide surprisingly amazing benefits for the body and brain, while other fats promote disease, and others are just neutral. All fats taste good, so why not choose to eat the most beneficial fats for your health and weight? After all, I'm not taking away fat from your diet. I'm just recommending a fat switch. Fix 9 is to replace less-healthy fats with healthier fats.

Once you know the perks of certain fats, I'm sure you'll want your family and especially the kids and grandkids to make the switch with you. Some of these good fats actually help you lose dangerous fat around your middle, while others help children's brain development and help keep your thinking skills from declining as you age.

So in fact, you get to have your fat and eat it too!

FAT FACTS

Let me set you straight on fats. We've spent decades conditioning everyone to think of fats as a bad thing and to follow low-fat diets. But it turns out that we were way off. Even though fat has more than twice as many calories as carbohydrates and proteins per gram, it's the type of fat that makes a big, fat difference to your health, not the number of calories. Certain types of fat are actually good for you, and despite the concentrated calories, those fats can help you lose body fat while boosting your health!

Unhealthy saturated fat, on the other hand, appears to increase chronic inflammation, including inflammation of the region of the brain that regulates hunger and metabolism.[1] These bad fats promote body fat and a need for more insulin, as well as a higher risk for disease.

If you replace unhealthy saturated fats (such as animal fat in meats and fats used in refined and processed foods) with good fats, such as olive oil and canola, you will be way ahead of the game.[2] The saturated fats found in dairy products appear to be more neutral, not as harmful as once thought, but not as helpful in improving health as omega-3 fatty acids and monounsaturated fats.

Two types of fat, omega-3 fatty acids and monounsaturated fats, can actually help your body to lower inflammation and burn calories. Even better, these healthy fats taste great and can be more satisfying than less-healthy fats!

OMEGA-3 FATTY ACIDS

You may have heard the term "omega-3 fats" thrown around. Let me fill you in since these omega-3 fats are so critical to your health. They are unsaturated fats that your body can't produce, and that tends to lower inflammation—the underlying cause of so many chronic diseases.[3]

Omega-3 fatty acids boost the health of your heart, immune system, and brain. In children, omega-3 fats support brain development to the extent that learning and behavior are impacted. Even comprehension, vocabulary, reading, and spelling skills can be impacted by omega-3 fatty acids.

Omega-3 fat types

The active forms of omega-3 fatty acids (DHA and EPA) are found almost exclusively in ocean algae and fish that have eaten algae, such as seaweed and salmon. However, eggs from barnyard chickens freely foraging for worms and bugs do contain some active omega-3 fatty acids as well.

The inactive and less effective form of omega-3 fatty acids (ALA) is found in some plant-based foods such as walnuts, flax, chia seeds, canola oil, hemp seeds, Brussels Sprouts, and omega-3 fatty acid-enriched commercial eggs. While ALA is still valuable, the body can convert only about 10 to 15 percent of it into the active form.

Because omega-3 fatty acids are not formed by the body, they must be obtained from food. It is essential that we eat them! Ideally, the active forms of omega-3s should come from the recommended two to three seafood meals or at least eight ounces of seafood (which can include canned salmon or tuna) each week. Omega-3 fatty acid supplements cannot replace the benefits of eating seafood.

FISH OIL SUPPLEMENTS

Don't replace actual seafood with fish oil supplements! While fish oil pills are helpful, recent research has indicated that the benefits of eating seafood exceed those of supplements. Plus, while fish oil supplements provide omega-3 fatty acids, they don't contain all the many less common yet critical nutrients found in seafood. The amount of fish needed for these supplements is negatively affecting the fish supply to a point that our future seafood supply and the oceans are at serious risk.

Algae-based DHA+EPA supplements, on the other hand, are a great way for vegans to get some active omega-3 fats in their diet without harming the oceans. If you are not eating fish two or three times a week, these supplements are a much more sustainable choice than fish-based supplements.

Omega-3 benefits

These powerful omega-3 fatty acids decrease chronic inflammation and seem to help fight inflammatory diseases such as arthritis and irritable bowel syndrome. They can also inhibit cancer and protect your heart.[4] Omega-3s reduce deaths from heart disease, especially sudden deaths. Some omega-3s protect against abnormal heart rhythms, lower the risk of blood clots, reduce triglycerides, decrease the growth of plaque on artery walls, and reduce the risk of a stroke.

Brain boost

Omega-3s protect brain function. They seem to be especially important for a well-functioning central nervous system and for the transmission of signals from the eyes to the brain. The omega-3 fatty acid DHA is integral to a baby's brain development and function as well as for the development of the retina in the eye and the vascular system.

Omega-3s are critical for learning, vision, and brain function.[5] Studies have indicated that breastfed babies with the highest omega-3 DHA levels have cognitive and IQ advantages.[6] Eating foods high in omega-3 fatty acids during pregnancy and nursing helps build babies' brains and eyes!

There is a reason why pregnant women are encouraged to eat seafood two to three times per week during pregnancy to boost brain development in babies. The World Health Organization makes the clear point that skipping seafood during pregnancy may mean missing out on the best possible brain development.[7]

Cognitive decline

The omega-3 fatty acids in seafood may also protect adult brains from decline.[8] In a study published in *Neurology* in 2016, age-related memory loss and thinking problems were greater in those participants who reported eating seafood less than once a week as compared to those who ate at least one seafood meal per week.[9] Research also shows that regularly eating seafood is linked to a lower risk of brain changes common in the development of Alzheimer's.[10]

Hints of memory loss shouldn't be your first signal to eat seafood high in active omega-3 fatty acids. Long-term intake of adequate DHA is linked to improved learning ability, improved memory, and reduced rates of cognitive decline. But to reap the brain benefits of DHA, you need to consume DHA-rich foods on a consistent basis.

The 2015-2020 Dietary Guidelines for Americans recommend that adults eat two to three servings per week or at least 8 ounces of seafood per week. Oily fish such as wild salmon, tuna, mackerel, herring, and trout provide plenty of active omega-3 fatty acid DHA to help preserve your aging brain.

BRAIN BENEFITS FROM THE SEA

Seafood has anti-inflammatory and nerve cell-protective effects that can reduce the risk of cognitive decline, depression, perinatal depression, and anxiety and can reduce the risk of developing Alzheimer's disease. Eating seafood food high in omega-3 fatty acids during pregnancy and nursing helps babies' brains and eyes develop! In children, omega-3 fats from seafood continue to support brain development to the extent that learning and behavior are improved.

Eat at least 8 ounces of sustainable seafood each week as part of an eating pattern that includes plenty of whole foods like fruits, vegetables, legumes, nuts and seeds, barnyard eggs, and healthy fats from avocados and olives to boost your brain health!

Mental health

Omega-3s are important for mental health too. Low omega-3 fatty acid DHA levels are associated with depression, memory loss, dementia, and visual problems.[11] Low

DHA levels during pregnancy can increase the mother's risk for depression during and after pregnancy.[12]

Childhood allergies

Inadequate intake of omega-3 fatty acids during pregnancy may increase the risk of food allergies and asthma in children. Studies indicate that omega-3 fatty acids consumed during pregnancy may reduce common food allergens and the prevalence and severity of atopic dermatitis in the first year of life (possibly until adolescence).[13]

Consuming omega-3 fatty acids during pregnancy may also reduce persistent wheezing and asthma in children between the ages of three to five years. Eating oily fish like salmon and light tuna along with algae omega-3 oil supplementation during pregnancy is well worth considering.

> **PLANT-BASED OMEGA-3 FAT FOODS**
>
> While seafood and sea vegetables are your best source of active omega-3 fatty acids (DHA and EPA), plant-based foods rich in the inactive ALA omega-3 fatty acid are important too. While the ALA omega-3 fats in food from plants need to be converted into the active forms (EPA and DHA) before your body can utilize them, don't miss an opportunity to eat more omega-3 fats and the other powerful nutrients that accompany these plant foods.
>
> Chia seeds are the best plant-based source of omega-3 fatty acids, even more than flax seeds. Walnuts, flax, chia seeds, canola oil, hemp seeds, and Brussels Sprouts also include ALA omega-3 fatty acids. Be sure to include at least some of these foods in your diet regularly.

Seafood surprise

News flash! Mercury toxicity is not an issue in most varieties of seafood, contrary to what was once suspected. Eating fish doesn't cause mercury toxicity; it actually prevents it! The key is the combination of mercury and selenium that people ingest, not the amount of mercury alone.

There is so much selenium in most seafood that it neutralizes the mercury, and you often end up with a surplus of selenium reserves.[14] This is good news because selenium recycles and restores many vital antioxidants, such as vitamin C, to their active form, protecting against damage in vulnerable tissues such as the brain. Sele-

COMMON FISH AND SEAFOOD PRODUCTS AND THEIR OMEGA-3 FATTY ACID CONTENT

FISH

39 mg
- 1 oz mahi-mahi

45 mg
- 1 oz Atlantic cod

51 mg
- 1 oz farmed catfish

60 mg
- 1 oz frozen fish sticks

76 mg
- 1 oz light Skipjack tuna

132 mg
- 1 oz halibut

134 mg
- 1 oz Alaskan pollock

235 mg
- 1 oz swordfish*

244 mg
- 1 oz albacore tuna**

264 mg
- 1 oz trout

278 mg
- 1 oz sardines

296 mg
- 1 oz wild salmon

600 mg
- 1 oz anchovies

751 mg
- 1 oz farmed salmon

SHELLFISH

24 mg
- 1 oz lobster

80 mg
- 1 oz clams

89 mg
- 1 oz shrimp

103 mg
- 1 oz scallops

117 mg
- 1 oz crab

154 mg
- 1 oz oysters

222 mg
- 1 oz mussels

*High-mercury fish to avoid: Swordfish, King mackerel, shark, marlin, orange roughy, and tilefish (sometimes called golden snapper or golden bass). Women who are or who may become pregnant, nursing mothers, and young children should avoid these high-mercury species of fish but should eat up to 12 ounces (two average meals) per week of a variety of lower-mercury seafood.

**Albacore tuna contains more mercury than canned light tuna. Everyone, but especially pregnant women, nursing mothers, and young children should choose canned light tuna over albacore tuna.

Mozaffarian D, Rimm EB. Fish Intake, Contaminants, and Human Health: Evaluating the Risks and the Benefits. *JAMA.* 2006;296(15):1885–1899. doi:10.1001/jama.296.15.1885.

nium also prevents and reverses oxidative damage throughout the body. All cells need selenium, and there is no substitute for it in the body.[15]

Seafood has other valuable nutrients too, such as active omega-3 fatty acids (DHA and EPA), choline, vitamin D, and iodine, which most people are not consuming sufficiently and which can be difficult to find in other foods. Pregnant women and young children are especially in need of these valuable nutrients but may be unaware of seafood's terrific health benefits.[16] For everyone, eating sustainable marine life is more important to your health than you can imagine, so it's time to get in the habit of buying it.

Omega-3 fats in eggs

Beneficial active omega-3s (DHA and EPA) are almost nonexistent in commercial eggs. There is a very small amount of inactive omega-3 fats (ALA) in your average industrial egg, but your body converts only about 10 to 15 percent of the inactive omega-3 fatty acid to the active form of omega-3s, which is so critical to your health and your brain.

Eggs from barnyard chickens that freely forage for worms and bugs much of the day, however, may contain as much as 120 mg of active DHA and 23 mg of active EPA per large egg, which is more than six times the amount provided by the eggs of grain-fed chickens.[17]

INDUSTRIALIZED EGGS

Several decades ago, when industrial eggs became the norm, people began to miss out on a significant source of active omega-3 DHA and EPA—local eggs. People who don't eat seafood (or who don't eat much tuna, salmon, and other oily fish) are particularly likely to be deficient in this health-promoting active omega-3 fat. In the late 1980s, researcher Dr. Artemis Simopoulos noticed when visiting Greece that the chickens were running around grazing in the wild fields, eating greens and insects. She smuggled some of those grass-fed Greek eggs back to the U.S. to do a fat analysis and compared them with the corn-fed eggs typically sold in American supermarkets.

The results were surprising: the omega-3 levels were far higher in the grass-fed eggs. More impressively, the omega-3 fatty acids were actually mostly active omega-3 fatty acids, like those found in seafood. Eggs from hens that forage in fields for bugs and worms are a good source of active omega-3 for non-seafood eaters.

MONOUNSATURATED FATS

Monounsaturated fatty acids, known as omega-9 fatty acids, are also "good fats." These fats decrease the bad cholesterol (LDL), increase good cholesterol (HDL), lower triglycerides, blood pressure, and also help control blood sugar levels.[18]

Exciting research suggests that monounsaturated fat may even help to reduce abdominal fat. Studies out of the University of Manitoba in Canada have shown that substituting monounsaturated fatty acids for saturated fats in your diet can help you lose belly fat.[19]

Monounsaturated fatty acids are in olive oil, refined (also known as high-heat) safflower oil, canola oil, avocados and avocado oil, peanuts and peanut oil, and tree nuts and other nut oils. These foods and oils are easy to integrate into your everyday meals and snacks.

> **WHOLE FOODS BEAT OILS**
>
> Whole foods in their whole-food forms are the best way to eat food! The thousands of plant chemicals, along with the fiber, vitamins, and minerals work together synergistically to improve your health. Choose avocados, olives, and nuts over using just their oils when possible.

Nuts and seeds tend to be high in monounsaturated fats or omega-3 fats. Macadamia nuts, hazelnuts, pecans, almonds, cashews, Brazil nuts, pistachio nuts, pine nuts, and sesame seeds are high in monounsaturated fats, while chia seeds, walnuts, flax, and hemp seeds are loaded with plant-based omega-3 fatty acids. It's no wonder that people who eat nuts tend to live longer!

Clearing up the canola oil confusion

There is a lot of confusion revolving around canola oil, so let me set the record straight. The plant is called canola rather than rapeseed. Canola oil was bred naturally from rapeseed in Canada in the early 1970s but has a different nutritional profile and contains less erucic acid than rapeseed. This misunderstanding has been the cause of canola oil getting a bad rap.[20]

Also high in monounsaturated fats (63 percent), canola oil has the added benefit of containing healthy omega-3 fatty acid (ALA). Try purchasing organic canola oil to avoid pesticides and genetically modified organisms (GMOs), but stick to the refined product, as it has a higher smoke point of around 400 degrees Fahrenheit. It doesn't have the polyphenols that are destroyed in heat like olive oil, and it has all the benefits of monounsaturated fat.

In fact, highly processed canola oil was studied in the Canola Oil Multi-Centre Intervention Trial, a large research trial conducted by several universities, which reported that monounsaturated fatty acid (MUFA) intake specifically reduced abdominal fat. After four weeks, the participants of this ongoing study lost 1.6 percent of their abdominal fat by consuming a smoothie high in MUFA along with a heart-healthy diet. The study also reported no rise in bad cholesterol (LDL) in the participants, confirming its cardioprotective attributes.[21]

Olive oil

Olive oil has plenty of well-documented health benefits because it's got a double whammy: not only is it rich in omega-9 fatty acids, also known as monounsaturated fats, but it's loaded with valuable plant chemicals that protect the plant from drought and disease. These health-benefiting phytochemicals are called polyphenols. They are associated with a lower risk of metabolic syndrome, reduced inflammation, reduced cell damage, and increased protection from infections. But not all olive oils are the same, and their valuable polyphenols are fragile and need to be protected from light, excessive heat, and age.[22]

The less processing the oil undergoes, the more nutrients it contains. Extra-virgin olive oil (EVOO) is pressed only one time without any heat, which is why it has the most polyphenols and the most health benefits. Adding heat to the olives helps producers to obtain more oil from each press, but the heat can damage the polyphenols and reduce the health perks.

The nutrient content of olive oil diminishes as the days, weeks, and months go by. Olive oil has a maximum shelf life of two years from the time of harvest. If the olive oil you're looking at is dated more than a year from its harvest date, don't buy it.

EXTRA EVOO HEALTH PERK

Extra virgin olive oil absorbs valuable polyphenols from vegetables during the cooking process. Researchers discovered that polyphenols are exchanged between veggies and olive oil when sauteing vegetables (like onions, celery, and carrots) on low heat. These phytochemicals, so fantastic for our health, are freed up and easier for the body to absorb after this low-heat preparation.[23]

FAT CHANCE

Foods high in trans fats escalate the amount of problematic LDL cholesterol in the blood and decrease the amount of advantageous HDL cholesterol. Trans fats cause inflammation that is linked to diabetes, heart disease, stroke, and other chronic problems. They keep your insulin from working as efficiently, so blood glucose levels are higher, and your risk of developing type 2 diabetes increases. Even little amounts of trans fats can be harmful. For every 2 percent of calories eaten each day from trans fats, the risk of heart disease increases by 23 percent.[24]

FDA'S TRANS FAT BAN

Although naturally-occurring trans fats from animal products are safe in moderate amounts, artificial trans fats or partially hydrogenated oils (PHOs) are harmful enough that their use in foods in the United States has been banned. The FDA's ban of artificial trans fats went into effect on June 18, 2018. However, products manufactured before June 18, 2018 have until January 1, 2021 to make the change.

It's processed food that's made with trans fats. So even when partially hydrogenated or hydrogenated vegetable oils are replaced by unhydrogenated oil or saturated fat, you still want to eat less processed foods with beneficial fats.

Limit commercially fried foods like french fries and processed foods like doughnuts, crackers, cookies, muffins, cakes, and pies. Make them at home with oils high in monounsaturated fats (and whole-food sweeteners instead of sugar). Replace foods made with saturated fats with nuts, avocados, fruit, popcorn made with monounsaturated oils, hummus, and more to benefit your health and weight!

When we are finally rid of trans fats in this country for good, less healthy saturated fats will still remain. Two other major studies have recently solidified the prescription to the saturated fat confusion. These studies clarified that replacing saturated fat with unsaturated fats (good fats in fish and from plants) as well as high-fiber, whole-grain carbohydrates is the best way to reduce the risk of heart disease. However, replacing saturated fat like bacon fat, fat in meat, and butter with highly processed carbohydrates causes disease.

Butter is not back

Recently the phrase "Butter is Back" has been making headlines. Despite this and similar articles selling the notion that saturated fat is good for us, it's not. The major study that set off these misleading articles was a 2014 review and analysis of twenty-seven trials and forty-nine observational studies from the University of Cambridge. Several researchers in the field took issue with the deep flaws in this study. Dr. Walter Willett, Harvard's Nutrition Department chair, was alarmed by several issues. The study left out results that conflicted with its conclusions and misrepresented some of its data. The study also neglected to consider the significant benefits of omega fats, so it missed the critical benefits of replacing saturated fats with healthier fats.[25]

Dr. David Katz from Yale explains that this study only showed that the rate of heart disease hasn't decreased in the US, despite our small decrease in saturated fat consumption. But Dr. Katz, like many other researchers, is quick to point out that it doesn't mean saturated fat isn't a problem. The real reason that heart disease didn't decrease is that Americans replaced saturated fat with sugar and refined grains rather than with vegetables and whole grains.[26]

Saturated fats aren't all the same, and they can vary in their effects on disease. Some saturated fats, like dairy fat, appear to be less problematic.[27] These saturated fats are more neutral than they are helpful, however. Spend your efforts in eating fats that improve your health and help you lose body fat.

Eating less saturated fat isn't enough. Replace butter, bacon, fatty meats, mayo, margarine, cream, and fried foods with more health-promoting fats in whole foods as often as possible. Whole foods like olives, avocados, nuts, seeds, and fatty fish contain healthy polyunsaturated and monounsaturated fats. Use avocado instead of mayo or nuts instead of croutons, for instance. When you need to cook with fat, go with olive oil or another plant-based oil instead of butter or bacon fat.

Coconut oil

While touted as a superfood, coconut oil just doesn't stand up to the excellent benefits of both omega-3 fatty acids and monounsaturated fats. Much of coconut oil's positive health contributions are attributed to its high content of medium-chain fatty acids (MCFAs), which tend to be absorbed and metabolized more efficiently than other fats and contain six to twelve carbons. Yet the medium-chain triglycerides (MCTs) (containing twelve carbons) in coconut oil act more like typical long-chain fatty acids even though structurally they are considered medium. Long-chain fatty acids need to be broken up and packaged in lipoprotein particles that transport fat around the body.[28]

While the traditional diets of Pacific Islanders historically included a lot of saturated fat from coconut products and the populations exhibited low incidences of cardiovascular disease, this is not sufficient to support coconut oil's superfoods status. These groups of islanders were eating grated coconut flesh, coconut cream, and coconut flour, not coconut oil. Unlike coconut oil, these coconut foods are rich in fiber, explaining in part why these groups had a lower rate of heart disease. Significantly, their traditional diets also contained plenty of fish, fruit, and vegetables and little to no refined sugar, processed foods, or soft drinks.

Other points supporting coconut's elevated health status also have similarly strong counterpoints, but I think you get the idea. Since about 70 percent of Americans believe coconut oil is a benefit to their health, it's important to clear up the confusion. Coconut oil can actually be bad for you. A 2020 Harvard meta analysis review of sixteen coconut oil studies found that coconut oil did not improve health, and a high intake of coconut oil increased the bad LDL cholesterol in the blood, leading to an increase risk of heart disease.[29] Plus, coconut oil isn't beneficial to your health like omega-3 fatty acids eaten in fish and monounsaturated fats eaten in nuts, seeds, olives, and avocados. I use unrefined coconut oil from time to time in some of my recipes for its taste and texture, but more often than not I use monounsaturated fats (olive oil, high-heat safflower oil, canola oil, avocados, and avocado oil, peanut butter and peanut oil, and tree nut and other nut oils) for real benefits.

LET'S GO SHOPPING!

Below are a few tips to help you purchase the best fats for your health and weight. Some of these fats come in your food, and some are oils on the shelf. But the best way to avoid less healthy fats is to not bring them home from the store! So skip the processed junk. If you want a treat, make it with the best oils, whole grains (chapter 10), and whole-food sweeteners (chapter 1). There are also some prepared foods that are healthier choices at the store. That's why you need to look at the ingredient list to know what's really in the product. The claims on the front of the label can be very deceiving. Happy fat shopping!

Spread the love

I know people want their peanut butter sweet and not too oily, but I promise, you will become acclimated. You will like healthier peanut butter if you stick with it for a while. Read the labels. Give brands that don't add extra sugar and fat (especially hydrogenated fat) a chance. Peanuts and maybe salt should be the only ingredients on the list. Yep, that's all folks!

Good old crushed peanuts are all you need to find real joy in peanut butter. Crazy Richard's 100% Peanuts is my favorite but any brand with just peanuts (or peanuts and salt) will work. You will be doing yourself and your kids a big favor because whatever they get accustomed to is what they'll eat for the rest of their lives. If you're having a hard time letting go of your favorite added-fat, sugar-laden peanut butter, then you know what I mean. You might as well help them get used to peanut butter that is healthier for them and just as enjoyable (if not more so) in the long run!

> **PEANUT BUTTER TRANSITION TIP**
>
> To transition from the popular sweetened brands (that will not be named), add a little raw honey or pure maple syrup to the peanut butter container. Over time, you can decrease the sweetener, and no one will notice.
>
> Mix the raw honey in when you stir the oil on the top into the peanut butter or with a hand-held mixer and one beater.

Nut butters

The same peanut butter shopping rules apply to nut butters; look for butters without added sugar and fat. You don't want anything added except the nuts and maybe salt. Try a variety of nut butters and even ones made with multiple types of nuts to see what you like.

If you want a Nutella-like experience but on the healthy side, try mixing cocoa and raw honey or pure maple syrup into your nut butter. You probably don't need to use a hand mixer for nut butters. Just stir those ingredients in. Enjoy!

> **PEANUT BUTTER OIL MIXER TIP**
>
> Read carefully! Take a hand-held mixer with just one beater in the mixer in one hand. Hold the peanut butter jar securely on the counter and hold the mixer with the other hand. Place the single beater (attached to the mixer) into the jar of peanut butter, but don't turn the mixer on until you are holding the jar with the other hand very tightly (I mean so tight that your life depends on it) and pressing it down on the counter. Turn on the beater to the very lowest setting to get the feel.
>
> Once you are feeling like you have a tight enough grip on the peanut butter jar, turn the mixer up until you have it on the highest setting. Whip that oil (and maybe raw honey) with the peanut butter until the oil is well mixed. To keep it mixed up for a long time, store upside down in the refrigerator. If you don't want your peanut butter cold, store it upside down in the cabinet. The oil doesn't separate for several weeks even without being refrigerated.

Buy nuts

Nuts can be expensive. But don't lose sight of all the nutrients and health-promoting fats you are purchasing in their whole-food package. It's premium quality food, so it's worth more to your body than overly processed foods that don't provide valuable nutrients.

Trader Joe's carries nut pieces that save you some money. They have a wide variety and are sold with or without roasting and added salt. Wholesalers, like Costco, allow you to buy larger bags of nuts to save on the cost too. Some stores and ethnic markets sell nuts in bulk bins that save you money on packaging. Nuts keep you more satisfied too, so you need less food overall. While nuts are expensive, they are worth the extra money!

WHAT ABOUT PEANUTS?

Technically, peanuts are a type of bean or legume, not a nut. Peanuts are similar to tree nuts nutritionally, but peanuts grow in the ground, and tree nuts grow on trees. Peanuts have a little more protein than the tree nuts, but peanuts contain an antioxidant called resveratrol (also found in grapes, wine, and soy), which may protect against heart disease, and arginine, which helps improve blood flow to the heart. As for shopping for peanuts—worry less about whether they are raw or roasted and more about buying lightly salted or even unsalted ones.

UNSCRAMBLE THE CODE TO BUYING EGGS

How do you select the "best" eggs for purchase when there are so many different nutritional claims on the egg cartons? It's enough to scramble anyone's brain! With so many mixed messages and so many claims to unscramble, buying eggs can be quite an undertaking! Let me break it down for you.

Best eggs

To get plenty of active omega-3 fats in your eggs, try finding eggs from a local farm. Farmers' markets are a terrific place to start, but ask questions to find out if the hens have plenty of time roaming around in the sun.

Egg carton claims decoded

If you end up in the grocery store to purchase eggs, look for egg cartons that claim pasture-raised eggs. "Pasture-raised" has no legal meaning, but that doesn't mean that the hens aren't outside scavenging for bugs. Look for information on the label that indicates how much land the hens get to roam on.

If the egg carton is stamped with a "Certified Humane Pasture-Raised" or "Animal Welfare Approved" on the carton, that will ensure that their hens are roaming around pecking away. If the egg carton is just stamped "Certified Humane Raised & Handled" without the word "pasture" included, you still need some indication on the label that the hens are really outside foraging often.[30]

An egg carton touting "cage-free" hens is a sign that the hens are not outside pecking at all. Cage-free hens are still stuffed in close quarters without real access to roam outside. Since these hens can be fed flax feed, a high omega-3 egg label also doesn't equate a with a better standard of living.

Free-range eggs have access to the outdoors, but how much and the quality of the area is unreliable. It's often in a small space without a field to peck at bugs and worms. Organic hens are uncaged and have some outdoor access, but they don't have foraging access since their diet is guaranteed to be an organic, all-vegetarian, and free of pesticides and antibiotics.[31]

SUPERMARKET OMEGA-3 FAT EGGS

You may be wondering about all the egg brands on the market that tout impressive omega-3 amounts in their eggs. Most often, those impressive omega-3 fatty acid numbers are from inactive omega-3 ALA fatty acids, not the active DHA and EPA omega-3s. These chickens are most often fed feed mixed with flaxseed to boost the egg's omega-3 values. Flaxseeds are healthy for us, but they don't contain active omega-3s DHA and EPA.

These hens would need to eat algae or fish oil to produce eggs with active omega-3s in them. In fact, consuming seafood and algae is the only other way to get the active omega-3 fatty acids, DHA and EPA in the eggs besides allowing hens to eat bugs and worms. That is precisely why eating eggs that contain active omega-3s is so critical to your family's health. Eggs from hens that forage for insects are also significantly high in other nutrients like vitamin E, D, and important B vitamins like folate and B-12.

GO FISH: BUYING FISH

Americans on average are eating only 10 to 20 percent of the two to three seafood meals (8 to 12 ounces) recommended per week. Many people and especially pregnant women have shied away from eating seafood because of mercury warnings.

It's easier to buy seafood now that the mercury toxicity issue is more clear. **Avoid** tilefish from the Gulf of Mexico, king mackerel, swordfish, and shark (which make up only 0.5 percent of total US fish intake). These are all species of fish that are low in selenium while high in mercury.

Ninety-seven percent of freshwater fish in the United States have proper balances of selenium and mercury, but lake- and river-caught fish are highly dependent on the selenium in the soil surrounding the watershed. Be sure to find out if the fish is safe to eat in a particular fresh body of water.

Purchase seafood (especially wild salmon, anchovies, mussels, rainbow trout, and sardines) that is particularly high in active omega-3 fatty acids (DHA and EPA). These inflammation-fighting fatty acids reduce the risk of cardiovascular disease and stroke as well as many other inflammation-related diseases, plus depression and anxiety. They may even help our minds stay focused.

OMEGA-3 FAT'S NEW LABEL ON SEAFOOD

As of June 2019, fish and seafood packaging can boast a qualified health claim stating that omega-3 fatty acids can help lower blood pressure and reduce the risk of coronary heart disease and hypertension. The labels will also provide the amounts of active EPA and DHA omega-3 fats.

Currently, no intake level of omega-3 fatty acids is recommended in the United States, but the Dietary Guidelines for Americans (DGA) do recommend that consumers eat eight ounces of seafood each week. That amount of seafood provides about 250 mg per day of EPA and DHA omega-3 fatty acids. But do note that this recommendation is for you to benefit from all the amazing nutrients that seafood provides, not just its omega-3 fatty acid content.

Seafood sustainability

Catching fish faster than they can reproduce is an urgent issue that is one of the biggest threats to our world's oceans. We want to eat more fish, not less, so we need to protect the fish that we have! Not all seafood is overfished, but we need to be aware of which species are in danger so we don't add to the problem when buying fish at the market or in a restaurant. Avoid bluefin tuna, wild sturgeon, all species of shark, orange roughy, skates and rays, most Chilean sea bass, most Atlantic halibut, and most marlin.

Farmed seafood

Farmed seafood or aquaculture is another source of confusion and has its issues. Some of these farm-raised fish cause concerns with regard to chemicals, disease and parasites, feed sourcing, escapes, and nutrient loading that can have severe environmental consequences if not monitored and carried out in accordance with set standards. However, the growing demand for seafood in the future can be met only through aquaculture. Fortunately, many of these fish farms continue to improve as third-party organizations have emerged and set a higher bar for the farmers. There are now some very sustainably farmed fish available, including salmon.

FARMED SEAFOOD THAT IMPROVES THE ENVIRONMENT

Some seafood farming actually helps the marine environment. Oysters, clams, mussels, and scallops tend to be sustainably farmed and actually filter the water to keep water cleaner and healthier. Other good bets include arctic char from around the world and US raised catfish and rainbow trout.

Finding the best seafood choices

Third-party groups, such as the Marine Stewardship Council, the Aquaculture Stewardship Council (ASC), and the Global Aquaculture Alliance Best Aquaculture Practices, help both seafood fishers and farmers stick to a high standard as they catch or farm seafood in an environmentally responsible way. They also help retailers meet their goal of sourcing all their seafood from suppliers with qualified eco-certification programs to provide sustainable seafood for you to buy. Make sure the seafood you buy has been approved by a third-party organization by asking at both stores and restaurants. Use the Seafood Watch app or FishWatch.gov website to look up your seafood choices. They are both easy to use, and you can search by type of fish when you are at a fish counter or ordering your dinner. Consumers' wants can help shape the ocean's health in big ways.

Also, branch out and try to choose a different fish variety. This helps to build a market for the abundant but underutilized amount of fish, such as scup, that is caught while fishers are seeking other types of seafood. Look for what's local and seasonal if you live near the coast. Take advantage of what's fresh in the marketplace rather than shopping for a specific type of fish.

HEALTHY, SUSTAINABLE SEAFOOD CHOICES

- Coldwater fish such as Atlantic mackerel, sardines, and herring
- Farmed fish such as US tilapia, US rainbow trout, US catfish, and US striped bass
- Farmed shellfish such as blue mussels, New England clams, US oysters, and US bay scallops
- Wild-caught fish, salmon, barramundi, US bluefish, Dungeness crab, and US Mahi Mahi

Avoid bluefin tuna, wild sturgeon, all species of shark, orange roughy, skates and rays, most Chilean sea bass, most Atlantic halibut, and most marlin. Also, avoid these high-mercury, low-selenium fish: tilefish from the Gulf of Mexico, king mackerel, swordfish, and shark.

More seafood shopping tips

Try to buy fish that is bright and translucent rather than dull. You also want fish that smells like fresh seawater, rather than fish that smells "fishy." It's completely fine and sometimes more convenient to buy frozen, sustainable seafood that is more likely to be very fresh.

When it comes to salmon, choose wild salmon from the Pacific, including both Alaska and Washington, which is a great choice for the environment and your health. Pouched or canned Pacific Ocean salmon is always available and easy to incorporate into meals, such as salads and main dishes. It's good for you, for the fish, and for your pocketbook.

EASY SALMON OR TUNA POUCHES

Purchase sustainable salmon or tuna with olive oil in pouches and take them with you to work, while traveling, or on errands. Eat it right out of the pouch or with whole-grain crackers (like Hint of Salt Triscuits) or on a green salad. What an easy way to help yourself eat the recommended two to three servings of seafood each week and enjoy a satisfying, high-protein lunch!

For those of you who are a little iffy about eating fish at all, you might try Barramundi, The Sustainable Seabass. It's a particularly delicious, mild fish that has the highest omega-3s of any white fish. You can buy it frozen in stores throughout the country. For stores in your area that carry Australis Barramundi, check out their website at www.thebetterfish.com.

Purchasing seafood doesn't have to be difficult or nerve-racking! You need to eat more of it to take advantage of its many health benefits. Use the Seafood Watch app when deciding what seafood to buy. It's so quick that you can find out the best choice while looking over the menu at a restaurant.

KITCHEN PREP RALLY

It's time to start prepping in the kitchen using these amazing health-promoting fats! Simple dressings for both salads and grain bowls are made with monounsaturated-rich oils and, in some cases, raw cashews. Use nuts, seeds, avocados, and olives in your recipes freely. Honestly, dressing and fish are quick and easy to prepare, so you won't be in the kitchen for long.

Using olive oil

Extra virgin olive oil is loaded with an extra bonus of polyphenols along with 73% percent monounsaturated fat and is my first choice for dressings where heat is not needed. Extra virgin olive oil can handle some heat (up to about 325°F) but heat does kill the valuable polyphenols responsible for many of its fantastic health benefits. When you are cooking over moderate heat to high heat, use light olive oil or olive oil that can withstand heat up to 468°F.

Cooking with olive oil is delicious, but there are a few tips you should know. When extra-virgin olive oil heats up over anywhere from 325°F to 375°F (depending on the manufacturer), it breaks down chemically and loses many of its phytonutrients. If the oil is heated beyond the breakdown temperature or smoke point, unpleasant and potentially unappetizing smoke is released.

Avocados

If you have a choice use the whole avocado rather than just the oil, to take advantage of the nearly twenty vitamins and nutrients including fiber, vitamins, minerals, and plant chemicals working together to help you feel full and boost your health. In fact, adding avocado to a vegetable salad actually helps your body to absorb more nutrients in the salad.

Blend avocado into a salad dressing rather than just adding the oil to thicken it; this way you can add even more body and creaminess to the finished product. Avocados can replace mayonnaise on a sandwich: cut open the avocado using a knife, and just spread the avocado on like a sandwich spread. This moistens the sandwich, and, of course, provides many more nutrients. You can't forget the famous avocado toast eaten plain or with eggs or vegetables like tomatoes. A little sprinkle of salt makes the avocado flavor come alive!

> **PEEL YOUR AVOCADO**
>
> The dark green outermost flesh of an avocado is the richest in powerful antioxidants, called carotenoid. Yet often that part gets thrown away. If you peel an avocado, you get more avocado and more nutrients. It's so easy to do if you cut the avocado in quarters instead of in halves lengthwise. The peel comes right off avocado quarters! Try it!

Nuts and seeds

Nuts in all forms can make your meal and snack recipes tastier and healthier! Add crunch to your food by adding nuts and seeds (like sunflower seeds) instead of overly processed foods. Replace croutons on your salad with nuts. Use plenty of nuts and seeds in your granola and to top oatmeal and yogurt. Make nuts a part of your snacks. Nuts are a nice addition to grains, pilafs, grain bowls, and baked goods.

Nut flour, like almond flour, adds moisture to a baked product. You can use the nut flour to reduce some (but not all) of the flour in baked recipes. Substitute the flour in a recipe for half nut flour and half either gluten-free flour or whole-grain flour. Even tortillas made with almond flour are being sold in stores throughout the country. And yes, they are delicious!

Chia seeds soften as they absorb liquid, so they add moisture and thicken your food. Use chia seeds to thicken sauces, especially darker color sauces to disguise the dark-colored seeds. Soak chia seeds in stock, wine, or juice and then add to your baked goods or ground meats like turkey balls. Plus, chia seeds add lots of healthy plant-based omega fats and many other nutrients including antioxidants that help preserve your food.

I add chia seeds to my berry fruit sauce to thicken it instead of spending the time to cook it down. The antioxidants in the chia seeds help the berry sauce last longer in the refrigerator. Usually, the blueberry sauce also gets used to top yogurt and to top healthier versions of pancakes, waffles, crepes, and french toast during the week. It's even a nice addition to ice cream on occasion. You might as well slip in health-promoting nutrients whenever you can!

MONOUNSATURATED-RICH HIGH HEAT OILS

Safflower oil *(78% monounsaturated fats, 450°F-510°F smoke point)* A nuetral oil that should not be heated, predominantly polyunsaturated fatty acid, and low in monounsaturated fats.

Canola oil *(63% monounsaturated fats, 400°F smoke point)* Contains plant-based omega-3 fatty acid (ALA). Use Organic canola oil to avoid pesticides and genetically modified organisms (GMOs) if you choose.

Avocado oil *(74% monounsaturated fats, unrefined = 480°F smoke point, refined = 520°F smoke point)*

Macadamia oil *(84% monounsaturated fats, 410°F-453°F smoke point)* A oil with minimal processing, so there is more nutrients.

Almond oil *(62% monounsaturated fats, 450°F smoke point)* Similar to other nut oils in make up and durability, but has a slightly higher smoke point.

Other Nut Oils: *(50% monounsaturated fats, 450°F smoke point)* Walnut oil and other tree nut oils are similar in makeup and durability. Walnut oil is fragile and should not be heated.

Peanut Oil *(48% monounsaturated fats, unrefined = 320°F smoke point, refined = 450°F smoke point)* Contains small amounts of resveratrol, a compound that may protect against cancer and heart disease.

Sesame Seed Oil *(45% monounsaturated fats, unrefined = 351°F smoke point, refined = 450°F smoke point)* Has a flavor that is too strong to be the main oil source of a recipe.

Olive Oil *(73% monounsaturated fats, refined = 390°F-468°F smoke point)* Extra virgin olive oil is unrefined with an extra bonus of polyphenols but has a lower smoke point of around of 325°F to 375°F depending on the manufacturer.

Hemp seed adds protein and other valuable nutrients to oatmeal, smoothie bowls, vegetables, and salads. Hemp seeds are nutrient-dense: 2 tablespoons provide 5 grams of protein, 2 grams of fiber, and good amounts of potassium and vitamin A as well as 25 percent of your iron needs. Try eating bell peppers or jicama slices with guacamole made with hemp and chia seeds. It's a real treat, and you may not even miss the chips.

CHIA SEEDS

Chia seeds are loaded with nutritious plant chemicals that thicken and naturally preserve food, so adding them to berry sauce makes it last in the refrigerator for more than a week. The seeds soften as they absorb liquid up to twelve times their weight. Because chia seeds are dark, they work especially well when making fruit sauce with darker fruit like blueberries. Make extra fruit syrup to add to plain yogurt and to top other whole-grain breakfast options like pancakes, waffles, french toast, oatmeal, and crepes. Chia seeds can be sprinkled on yogurt, cereal, smoothies, and oatmeal and added to baked goods like quick bread and muffins.

FISH COOKING IDEAS

Many people don't eat fish because they aren't sure how to prepare it. Don't let cooking fish scare you! It's so easy, and it can be much faster than chicken, pork, or beef. Here are five simple and delicious ways to prepare fish.

Broil your fish

This method is easy, fast, and tasty. Turn the broiler in your oven to high and cover a baking sheet with foil. Place your fish on the sheet, brush it with high-heat olive oil, add salt and pepper or a thin layer of miso paste or Dijon mustard, and slide the sheet pan into the oven so that it's about four inches below the heating element. The general rule of thumb is to cook the fish eight to ten minutes per inch of thickness (for quarter-inch fillets, it would be two to four minutes). When the fish is opaque and a knife slides easily into the thickest part of the fillet, it's done. Make sure you don't overcook it! Charred chopped tomatoes, ribboned summer squash, and red onion can cook with the fish on the same pan. Add capers to this vegetable mix and eat over the fish.

Roast your fish

Roasting fish is similar to broiling it, but it takes a few minutes longer. Place the fish and a little oil on a hot, preheated pan or baking sheet and put it into a 450°F oven to cook. Salmon is good roasted in the oven with just a little olive oil, salt, and pepper, but you can also use any fresh herb mixture with reduced-salt tamari or soy sauce and real maple syrup, cooked down, reduced to a thicker, sauce and brushed onto the salmon. Make your meal even easier by wrapping vegetables and fish together in foil.

Pan cook your fish

You can pan-cook fish in less than fifteen minutes. For thin fillets, try dredging the fish in sesame seeds or crushed nuts on both sides. Add to a large nonstick skillet over medium-high heat with one to two teaspoons each of sesame seed oil and a neutral oil such as canola. Cook for about two to three minutes or until seeds or nuts on the bottom are toasted and the top starts to turn opaque. Then cook on the other side one to two minutes. Add reduced-sodium soy sauce (with the sesame seeds) or salt and pepper. Keep in mind that the stovetop works best for fish that are on the thinner side, about a quarter-inch thick or less, unless you take a tip from many restaurants that sear their fish in a pan on both sides and then place it in the oven to finish the cooking.

Steam your fish

Try steaming your fish on top of vegetables. Start by sautéing garlic, onion, and some vegetables of your choice in a large skillet. Then add your fish—sprinkled with salt and pepper—onto the vegetables five to ten minutes before they're finished cooking (depending on how thick your fish is). Add some liquid (such as a can of diced tomatoes and its juice), adjust the heat to a simmer, and cover the whole pan to steam the fish until it is opaque and a knife can easily cut into it. Finish the vegetables, stirring occasionally. Season with fresh herbs and salt and pepper to taste.

Grill your fish

Grilling fish is another mouthwatering option, particularly in the summer, or just on an indoor grill. Heat your grill until it is very hot, above 400°F, and mist or rub both sides of the fish with a high-heat oil and your favorite herbs and spices. Even using just oil with salt, pepper, and lemon tastes terrific. If there is skin on the fish, keep it on to help hold the fish together.

Sear each side. Continue cooking until the center's temperature reaches between 130°F and 135°F; it will continue to heat up to around 140°F while it rests. If you like your salmon on the rare side, take it off at 120°F. Don't overcook it; the fish will dry out. For a fail-safe way to cook fish on the grill, wrap it in foil, alone or with vegetables. Don't forget to squeeze a little lemon onto the fish before serving, and savor every bite!

FIX 9

SWITCH FAT

Replace less-healthy fats with healthier fats.

For many years, fat was considered bad for our health. However, scientific research has since shown that many kinds of fat are extremely beneficial for our weight, health, mental health, memory, and thinking skills. Eating seafood two to three times per week is especially important for pregnant and nursing women for their developing babies. Children should also be eating plenty of seafood each week to support their developing brains. Adults need seafood to help prevent cognitive decline that comes with aging, along with many other health perks.

Fats with benefits

All fats taste delicious, so why not eat fats that benefit you in such amazing ways instead of eating fats that negatively affect your health and leave you feeling lethargic? Replace your meat with grilled seafood; replace lard, butter, shortening, and margarine with olive oil and canola oil; replace mayonnaise with avocados; and replace sour cream with plain Greek yogurt. Use oils high in monounsaturated fats like olive oil, canola oil, high-heat safflower oil, avocado oil, and peanut oil, as well as nut oils.

Switch it up

Eat raw nuts and avocados on salads and for snacks instead of processed foods to instantly increase your intake of healthy fats. Try to eat seafood, especially wild Pacific salmon, sardines, anchovies, tuna, and trout, at least twice per week to consume more omega-3 fatty acids in their active whole form. Flax, walnuts, chia seeds, and canola oil are also high in omega-3 fatty acids, although in a much less potent form that the body needs to convert into the active DHA and EPA form.

If you can't replace all your less-healthy fats with good ones right away, don't stress. Just work on gradually replacing questionable ones with healthier ones. Nothing in this plan is all or nothing. Every change you make will help you over time—even if it's small!

Action Plan

- ✓ Use extra-virgin olive oil for salad dressings but higher-heat oils such as light olive oil, canola oil, avocado oil, or refined safflower oil for cooking with high heat.
- ✓ Replace butter in baking with canola oil, refined safflower oil, or nut butters.
- ✓ Replace sour cream with plain Greek yogurt.
- ✓ Replace fried foods with fruits, vegetables, grilled fish, etc.
- ✓ Skip store-bought salad dressings and mix one part olive oil with one part white balsamic vinegar. Add salt and pepper and a little Dijon mustard. Shake vigorously for an easy dressing.
- ✓ Replace mayonnaise in sandwiches and other foods with avocado that is lightly salted, in slices or as a spread.
- ✓ Buy eggs from hens that are outside foraging for bugs and insects if possible.
- ✓ Eat healthy, sustainable seafood at least twice per week. Prepare with olive oil or avocado oil instead of butter.
- ✓ Replace butter with olive oil, refined safflower oil, canola oil, or avocado oil in cooking.
- ✓ Add avocado to green salads to increase your nutrient intake.
- ✓ Replace croutons in your salad with nuts and seeds. Eat raw nuts for a healthy snack.
- ✓ Buy peanut butter or nut butters without added fat or sugar. Transition from sweetened peanut butter by adding a little raw honey to the container. Reduce the sweetener over time.
- ✓ Track your progress on the *12 Fixes to Healthy* app.

FIX 10
CHANGE GRAINS

Carbs have gotten a bad rap, but don't throw all carbohydrates into the same camp. You may not realize how critical whole grains are in helping to feed the health-promoting bacteria in your gut and to help you drop fat around your middle. You can even lower your risk of disease by switching out refined grains to whole-grain foods!

Whole, ancient grains are a breeze to make. So it's time to pull yourself out of the potato, pasta, and rice rut and to expand your whole-grain horizons.

Make the switch and replace your refined grains like white, refined bread and buns, white rice, and white flour pasta with whole grains like whole-grain bread, black or brown rice, and whole-grain pasta. Your taste buds will rejoice, while your body does the happy dance!

Fix 10 is to eat between three and six servings of whole grains per day instead of refined carbohydrates.

WHOLE GRAINS FOR MORE DAILY DETOX

Dietary fiber found in whole grains (such as oats, brown or black rice, sorghum, farro, quinoa, barley, popcorn, and whole wheat), fruits, vegetables, and legumes are ranked among the strongest components of food to fight chronic inflammation in the Dietary Inflammatory Index.[1] It appears that the fiber ferments during digestion and produces a fatty acid that blocks inflammation. This type of fiber is particularly high in whole grains, dry beans, and peas.

Carbs differ

Carbs, and wheat in particular, have been demonized. Sugar and refined flour (typically white flour) deserve their bad rap. These processed carbs are associated with obesity, chronic disease, mental decline,[2] and depression.[3] But whole grains, which haven't been refined and stripped of valuable nutrients and bulk, are haphazardly thrown into the bad pot. People seem to fear all carbs without knowing the health value they can obtain from eating a few whole grains each day.

Whole grain benefits

There is a plethora of research revealing the many health benefits of whole grains.[4] Beyond helping to control blood sugar, these undefiled grains lower the risk of type 2 diabetes, heart disease,[5] high blood pressure, and cancer. Not only do whole grains help control weight but eating them instead of refined, processed grains cuts down on the amount of body fat around your middle.[6]

Whole grain defined

A whole grain is just that: whole with all parts of the grain together, even when it has been ground into flour or made into a wet mash. Whole wheat includes three components: the bran, the germ (chaff), and the endosperm (mostly protein and starch). One hundred percent whole wheat flour contains all three of these components in the same proportion as in the original grains, while white, refined flour is mostly endosperm without the fiber and nutrient-rich germ and bran. Some of the lost nutrients are replaced with a synthetic version of the nutrients that are absorbed less efficiently by the body. The nutrients niacin, riboflavin, thiamin, folic acid, and iron are required by law to be replaced, but the other nutrients lost, including fiber, are not replaced.

Steady fuel

We need health-benefiting whole grains for longer-lasting energy fuel, for both our bodies and our brains. Our muscles need to keep stored energy so we are energized to move. Our brains are more mentally alert and focused throughout the day when we eat whole grains. The fiber and phytonutrients in whole grains help to release glucose slowly into the bloodstream, providing a steady supply of energy to the body and brain.[7]

The fiber in whole foods is especially effective in slowing down the absorption of sugar and keeping insulin from spiking. Those who eat more fiber gain less weight over time.[8] So, up your fiber intake by eating more fruits and vegetables, lentils, beans, ancient grains, whole-grain bread, nuts, and seeds. If you are a woman, try to aim for at least 25 grams of fiber per day—the amount found in about four servings of whole grains and about five to six servings of produce. Men need more: shoot for 38 grams of fiber each day.

FIBER-FILLED DAY

Eating fiber is a critical boost to your gut, weight, and health. If you are a woman, try to eat at least 25 grams of fiber per day. Men should eat around 38 grams of fiber each day. You don't need to keep track of your fiber, however; eating at least three servings of whole grains per day instead of processed carbohydrates and half your foods in produce will cover your fiber needs, no problem!

This is what a fiber-filled day (33 grams of fiber total) looks like:

- half a cup of rolled oats (4g)
- half a cup of blueberries (2g)
- one ounce of almonds (4g)
- two pieces of whole-grain bread (4g)
- half a cup of black rice (3g)
- one cup of spinach (4g)
- half a cup of black beans (7.5g)
- a quarter of an avocado (4.5)

Gut boost

Fiber and resistant starch found in whole grains help digestion. The fiber keeps your bowels moving and helps ward off colon issues such as constipation and diarrhea, among other problems. The bottom line is to replace bread, rolls, crackers, and anything made from refined, white flour with intact whole grain as often as possible to keep your gut running smoothly.

The three pounds of microbes inside the digestive system include around some forty trillion microbes and the microbiome, is now considered a living organ. These little bugs are working hard to absorb nutrients from food, squeeze out invaders, and boost your immune system against short-term sickness and long-term disease. The gut microbiome is the balance of these microbes in your gut.

Whole grains are an important food for the health-promoting microbes in the gut, and they increase good bacteria in the large intestine for better digestion, greater absorption of nutrients, and a stronger immune system. Refined carbs contribute to an imbalance of microbiota in your gut that can negatively affect your weight, ability to fight off sickness, risk of disease, and even your responses to the world around you. Switch your grains to whole grains; your gut depends on them.[9]

Burn more body fat

Our bodies react differently to different types of calories. Foods from refined carbohydrates such as white flour and sugar slow down your ability to burn calories. They tend to cause your body to make more fat and make you hungrier more quickly, craving more refined carbs.[10]

Beyond the thousands of nutrients gained from whole foods, studies indicate that whole-grain foods burn more calories than processed grains. It takes energy to digest food. People use about 10 percent of their calories digesting and absorbing food, and eating food closer to Mother Nature's original state seems to take more energy to process.[11]

Whole foods burn more calories

In a study out of Pomona College in 2010, researchers fed two types of cheese sandwiches to different control groups. One group received sandwiches made with white bread and processed cheese. The other was fed sandwiches that consisted of whole-grain bread and cheddar cheese.

Both cheese sandwiches had the same number of calories to begin with. However, the researchers found that participants who ate the processed sandwiches used 47 percent less energy to digest their food. In other words, those who ate the whole-foods sandwiches burned almost twice as many calories as their processed-food counterparts did. Equally important is that six hours after eating the refined sandwiches, study subjects burned fewer calories at rest than the whole-foods group.[12]

Whole grains burn more calories

Another study recently published in the *American Journal of Clinical Nutrition* suggested that eating whole grains instead of refined grains burns more calories and increases metabolism. For the first two weeks of this tightly controlled study, 81 men and women ages 40 to 65 ate the same foods, including grains, in amounts that maintained their individual weights. For the following six weeks, they ate nearly identical foods at previous calorie levels, except their grains were either whole or refined grains.

Those designated the whole-grain foods absorbed fewer calories and burned more calories at rest. In fact, whole-grain eaters burned up to one hundred more calories per day than those assigned to eat refined grains.[13]

When sugars and starches are stripped of their fiber, the body doesn't have to work as hard to break down and absorb them. And over time, excess calories from processed foods add up fast. Eating whole grains burns more calories, leaving fewer of them to be stored as fat.

WHOLE GRAINS FOR LESS BELLY FAT

Fat around your midsection is a strong indicator of your health. Eating whole-grain-rich foods appears to help keep abdominal fat down. Participants of a study eating whole grains had greater loss of belly fat than those in the study eating refined grain foods, even though overall weight loss was comparable. The participants who ate the whole grains also had less inflammation.[14]

Live longer

Want to statistically reduce your risk of premature death from all causes by 15 percent just by making one change to your diet? Choosing to eat whole grains instead of refined-grain foods may do just that. After analyzing data from more than fifteen thousand people aged 45 to 65, researchers from the University of Minnesota School of Public Health reported that as whole-grain intake went up, the rate of death from all causes went down.[15]

Another analysis of data from fourteen studies with 788,076 participants reported that those who ate the most whole grains had a reduced risk of all-cause death by 16 percent and a reduced risk of cardiovascular-related death by 18 percent.[16]

HOW MANY WHOLE GRAIN SERVINGS?

One serving of grain (like rice and quinoa) is usually half a cup of grains or one slice of bread. Two slices of extra-thin sliced bread count as one serving, while airy popcorn contains three cups in one serving. Keep track of the number of slices and servings of grains you eat each day.

Lose inches

Try to eat almost all your grains as whole grains. If you want to lose inches or keep those inches in check, stay within the range of three to six whole-grain servings per day.

More than six servings of whole grains can be too much if you want to lose inches. For those engaging in moderate exercise less than an hour per day and who are of average to small-build, the three to six serving range should be about right. If you are a smaller build female exercising less than an hour per day and you want to lose inches faster, stay closer to three servings of whole grains and skip refined grain altogether 95 percent of the time. But don't be tempted to skip the minimum amount of three grains servings per day for maximum ability to lose body fat.

Whole grains for fuel

If you're doing intense exercise, of course, you may need more than six whole grain servings. If you are a larger person or are in athletic training, you will almost certainly need more than six servings of whole grains. If you aren't in training but exercise strenuously for more than sixty minutes on any given day, you still may need to increase your servings of whole grains that particular day to provide fuel for your extended workout.

Refined carbs

Put the emphasis on whole ancient grains and oats, but be sure to track all the whole grains you eat. Of course, when you eat refined, processed grains, such as white bread or white rice, rather than a whole grain on occasion, the refined grain should be counted as part of the three to six grains you eat each day if you are trying to lose weight. But do know that you are replacing foods that actually help you to decrease your body fat, to absorb fewer calories than refined carbs, to fight disease, and to keep your mind alert. Instead, you are consuming something that increases your body fat and contributes to illness. However, you should allow for a few guilt-free refined grains each week to give yourself some flexibility.

Better carb eating plan

Even though three to six servings of whole grains may be less than what you're used to eating, this is not a low-carbohydrate plan. It's a better carb plan! Fix 4 of these 12 Fixes encourages you to eat half of your meals and snacks in fruits and vegetables, which are also great sources of fiber-rich carbohydrates. Vegetables such as beans and lentils as well as starchy vegetables are very filling. Also, whole grains are more satisfying than processed grains. So you can count your grains and not worry about going hungry too soon!

A WHOLE GRAIN FOR EVERYONE

While whole grains are healthy for most people, they may not be tolerated well by all people at all times. Those with Celiac disease and gluten sensitivity should not eat gluten, a type of protein that some people are intolerant or allergic to. Wheat, barley, and rye contain gluten. But many grains do not, so there are many other grains free of gluten to choose from. Other digestive issues like irritable bowel syndrome may also warrant a search for a better tolerated whole grain as well.

Gluten-free whole grains, including rice, sorghum, buckwheat, oats, quinoa, and amaranth may be a solution for those with these conditions. Oats don't contain gluten but are made in facilities that contaminate the oats with gluten, so purchase oats that are clearly labeled "gluten-free."

Even those with digestive issues can often find some grains they tolerate. In fact, all grains are easier to digest when they are either sprouted or fermented.

Sprouted grains

Sprouted grains straddle the line between a seed and a new plant. Catching the sprouts during this germinating process breaks down some of the carbohydrates to make the grains easier to digest than regular grains. This process breaks down chemicals like lectins and phytate (a form of phytic acid) that typically decreases the absorption of nutrients, and sprouting increases the availability of almost all nutrients (including folate, iron, vitamin C, zinc, magnesium, and protein). Various phytochemicals (antioxidants) occur at higher concentrations.[17]

Fermented grains

Friendly microbes feed on the carbohydrate of the grains to produce lactic acid in another fermentation process. Whole grain sourdough bread is a good example. A starter full of good microbes consumes the carbohydrate portion of the flour to rise the bread. The result is fewer carbohydrates and more sour-tasting bread. This process naturally neutralizes those plant chemicals (like lectins and phytic acid) that prevent the absorption of nutrients. The nutrient content of your grains is greater, and they are easier to digest and provide better blood sugar control as well.[18]

Despite difficulty tolerating whole grains for some, there is most likely a whole grain for everyone to enjoy. Both sprouting and fermenting grains help those who are sensitive to digesting grains and allows them to include whole grains in their diet and to take advantage of their great health benefits.

COMMERCIAL WHOLE WHEAT FLOUR SURPRISE

Commercial whole wheat flour is milled by separating the endosperm from the germ and bran, similar to milling white refined flour. Unlike white flour, the germ and bran are added back in order for it to be called whole wheat. The FDA mandates that anything labeled "100 percent whole grain" must contain all three components in proportion to the original wheat seed. However, it doesn't dictate that the whole grain remains together during milling, as is done when you simply grind wheat into flour at home. The FDA allows manufacturers to make statements such as "100 percent whole grain" but not that the food is an "excellent source."

Most of the research reporting the health benefits of whole grain use this commercially milled whole wheat flour, so there are major benefits to eating it over bread made with white, refined flour. However, eating whole grain bread made with flour where the components were not separated and added back in may even be healthier. Certainly, sprouting whole grains or using microbes with a starter to make bread is more nutritious than commercial whole grain bread, which in turn is still better than refined, white bread.

FUEL FOR FITNESS

Fuel your engine! Your body can't run on fumes for long. Your primary focus before exercising should be on filling up your gas tank and making sure that you're hydrated. Carbohydrates are converted into fuel for your body to run.

If you haven't eaten all night, at least have a little snack of yogurt or bananas with peanut butter before you exercise. Then back up your workout to a meal within a couple of hours after ending your exercising.

Within thirty minutes before working out, eat some simple carbs in whole foods like fruit and dairy (yogurt, banana, or an apple). You don't always have to resort to eating and drinking refined, processed carbs to consume simple carbohydrates.

Complex carbohydrates like beans, lentils, whole grains, and starchy vegetables will provide long-lasting fuel for exercise along with so many nutrients and fiber. Unlike refined carbs, like white bread and buns, white flour crackers and cookies, soda, and many prepackaged processed foods, complex carbs found in whole foods won't supercharge your cravings for processed high sugar or salty foods. Steer clear of overeating right before you work out. Stop eating whole-grain carbs two to three hours before your workout.

BUY GREAT GRAINS

For more alternatives for your three to six servings of whole grains per day, choose ancient grains such as farro, sorghum, quinoa, spelt, Kamut, freekeh, millet, amaranth, teff, einkorn, and emmer when possible. They are more widely sold in typical grocery stores now, as well as Whole Foods-type stores, and can often be found in wholesale stores such as Costco for a reasonable price. These whole grains can also be found in the bulk section and shelves of stores geared to health. Some ethnic stores carry whole grains traditionally used in their cuisine.

Whole grain label lingo

Labels with terms such as "multigrain," "100% wheat," "organic," "pumpernickel," "bran," "wheat flour," and "stone-ground" may sound healthy, but technically none of these words and phrases means that the product is actually whole grain.

For example, a label claiming that the product in question is multigrain just means that the product contains more than one type of grain. That can easily mean that it has more than one type of refined grain.

By the same token, "100% wheat bread" means only that the bread is made from wheat flour. That flour may be refined—meaning that it has been stripped of the bran and germ—and the statement would still be true.

When you buy grains, look for the word "whole." "Whole" is the magic word here. Check the ingredient list. If it reads "whole wheat flour" or "whole" whatever type of flour, then you're in good shape. It's best to have whole grain be the first ingredient on the list, meaning that the product contains more whole grain than any other ingredients.

Sprouted grains

Sprouted grain products are popping up in grocery stores. When whole grains and seeds are soaked and left to germinate and the new sprout is still shorter than the original grain, then it is considered a sprouted grain. Sprouting grains increases many of the nutrients, including the B vitamins, folate, fiber, vitamin C, and essential amino acids that are often lacking in grains (such as lysine[19]).

At this point, there is no regulated standard definition of "sprouted grains," so sprouted products could vary. The nutrient changes that occur with sprouting differ greatly depending on the type of seed sprouted. There are also so many other variables that go into sprouting grains or seeds that you can't know under what conditions or for how long the sprouting has occurred in these products.

Labels shouting about sprouted grains or seeds may not be a good indicator of how much of the sprouted grain is used. Processed foods may use sprouted grain or seed as a way to lure you into thinking they are healthy. Is the product also made with refined flours? Even ancient grains can be made into refined flour. Look at the ingredient list and evaluate what flours are used and where they are on the list. Look for the magic word: "whole." The higher up on the ingredient list, the more there is of that ingredient. What other ingredients are used, what type, and how far up the list is sugar?

SPROUTED WHOLE-GRAIN BREAD

I recently invited myself to the Angelic Bakehouse plant in Milwaukee because I wanted to see for myself why these sprouted whole-grain products tasted better than other brands that I had tasted previously. I was impressed.

Angelic Bakehouse uses seven freshly sprouted whole grains instead of sprouted flour to make all their sprouted bread products. They sprout quinoa, red wheat berries, oat groats, millet, amaranth, rye berries, and barley for twenty-four hours right there in their kitchen to make all the bread, flatbread wraps, pizza crust, buns, rolls, and bread crips, as they've done since 1969.

Instead of using flour, they're using a freshly made wet mash of these seven sprouted whole grains so that their bread and bread products are more nutritious, naturally sweeter, moister, and tastier than other sprouted bread using sprouted grain flour. No preservatives like BHT are used, either; they use a little bit of prune juice instead for their preservative. Their breads taste better and are moister than other sprouted bread brands that use sprouted flour instead of a fresh wet mash of sprouted grains.

Whole-grain sourdough bread

You might be wondering which breads are the best for you. Ideally, whole grain sourdough bread made from a starter full of nutritious microbes, taking three days to make instead of three hours like commercially made bread, is a superior choice. In the bread-baking world, that's referred to as long fermentation.

Whole-grain sourdough bread is trendy, and if you can find someone with a good starter to give you, you can make it yourself. That may not be feasible for many, but whole-grain sourdough bread is popping up at good bakeries around the country. Make sure it's made with long fermentation time and is made with mostly whole grains.

WHOLE-GRAIN SOURDOUGH BREAD BUYING TIP

Buy whole-grain sourdough bread from a local bakery and have them slice the bread. Keep it in the freezer, and take slices out as you need them. Throw a frozen slice in the toaster on the highest setting, and you have amazing toast in minutes. Or set the toaster to medium-low and remove the bread when it is thawed and warm but not toasted. Single slices don't take long to thaw, but you can take out more than one slice in the morning so that the slices are ready to go during the day. This freezer storage method allows me to have fresh whole-grain sourdough bread when I need it, without much time invested.

Whole-grain flour options

When it's time to make muffins, pancakes, bread, or even cookies, you may wonder which whole-grain flours can be used to replace refined, all-purpose flour. Many gluten-free options are also available at the store. If you want to use a wheat-based flour for your recipe, let me suggest three wheat flour options: sprouted wheat flour, freshly ground whole wheat flour, or white whole wheat flour.

Sprouted wheat flour

Sprouted whole wheat flour is available at stores and is a great option to use in your baked products. I often keep this sprouted flour in the freezer, and it's a convenient option when I bake.

Freshly ground whole wheat flour

Freshly ground wheat flour tastes so good and is tastier in baked products (like pancakes) than grocery store whole wheat flour. Do a taste comparison using grocery store whole wheat flour and freshly ground if you get the chance. Many don't realize that grocery store whole wheat flour is milled similar to white flour but all the original parts of the flour are added back to the flour. If you grind your own wheat flour, you know it's intact from the beginning. This may sound daunting, but it's actually just another step and a wheat grinder. No worries, however. There is another option.

White whole wheat

Besides these two flour options, try white whole wheat. It sounds like a contradiction, but it's not. The name can be confused with refined, all-purpose white flour, but it is different. This white whole wheat flour tastes more like white, refined all-purpose flour but is nutritionally similar to standard whole wheat flour.

Unlike regular refined white flour stripped of the germ and the bran (full of nutrients and fiber), white whole wheat is milled from hard white wheat berries giving it a milder flavor and lighter color. Typical whole wheat flour is milled from hard red wheat berries, so it has a stronger flavor. White whole wheat tastes and bakes more like the refined all-purpose flour called for in most recipes but includes the germ, bran, and endosperm from white wheat. It's just another healthier option than white refined flour.

OVERLY PROCESSED FOOD POSING AS WHOLE FOOD

Labels and health slogans can be misleading—especially when it comes to overly processed foods and convenience-store snacks. That nutritionally hyped-up granola or power bar that you like doesn't compare to Mother Nature's real food in its whole form.

What they don't tell you is that the makers of highly processed foods often add overly refined flours, fibers, and proteins to increase the values on their nutrition labels. Those numbers support the superficial claims they make, but they don't often support your health as whole foods do.

For example, many cereal manufacturers are anxious to increase the fiber content of their cereals—but they don't want to alter the original recipe. So they extract fiber from whole grains and add it into the mix. Unfortunately, that fiber isn't nearly as effective without the whole grain itself; in their natural state, thousands of nutrients work together to create powerful health benefits.

Granola and protein bars

When you're grocery shopping, beware of foods that use extracts of certain nutrients. The longer the ingredient list is on the label, the more probable it is that the manufacturer is using pieces from whole foods to substitute for whole foods themselves. Granola, protein, and sports bars manufacturers often use this trick to raise the protein and fiber numbers on their nutrition labels. Try to choose brands that pack nutrients into their bars with natural methods, like certain types of Kind bars that contain the whole-food goodness of nuts that have only five grams of sugar or less.

If you're used to buying many overly processed foods, it can be hard to break the habit. Stick with it! After you avoid highly processed foods for just a couple of months, it gets a lot easier to keep bad shopping choices out of your grocery cart. Your taste buds adjust to eating healthy foods, and the processed foods you used to enjoy are no longer as appealing as they once were.

ARTISAN BREAD

Artisanal bread starts with just four ingredients—flour, salt, water, and yeast—and bakes into crusty, chewy loaves so fragrant and delicious that you cannot stop eating them. Try to find artisan bread with flour that includes whole grains! Avoid commercial bread with added stabilizers, preservatives, dough softeners, and added sweeteners to cover up the chemical tastes of these additives. Bread made with twenty-five or more ingredients is not unusual.

OATS

Oats deserve your praise and attention. They are prized for their nutritional value and health benefits. The Food and Drug Administration even allows whole-grain oats to use a health claim on food labels in regard to a reduced risk of coronary heart disease.

All the health benefits of whole grains apply to oats, but oats stand on their own too. The primary type of soluble fiber in oats is beta-glucan, which helps to slow digestion, suppress appetite, and increase satiety. This special Beta-glucan in oats can bind with cholesterol-rich bile acids in the intestine and transport them through the digestive tract and out of the body. Whole oats contain plant chemicals that reduce the damaging effects of chronic inflammation associated with various chronic diseases, like cardiovascular disease and diabetes.[20]

Include oats as part of your whole grain servings regularly for their many health perks. The least processed oats, like groats or steel-cut oats, usually take longer to digest, so they do a better job at keeping blood glucose levels steady even though nutritionally speaking they are similar to other oats. Rolled oats (old-fashioned) and Scottish oats are nothing to look down on, however. All these forms of oats are satisfying and add a great deal of health-promoting benefits to your diet.

PREPARING GREAT GRAINS

Don't let preparing whole grains intimidate you! It's easier than you think. Start with oats for breakfast.

Quick or instant oats are overly processed compared with good old rolled oats. To make instant oats, groats are steamed for a longer period of time and rolled into smaller pieces so that they can absorb water easily and cook very quickly. Most brands of instant oats come sweetened and flavored, so choose rolled oats or other forms over these instant oats. At the very least, check the ingredients for no added sugar.

QUICK OATMEAL FROM ROLLED OATS

For an easy and fast oatmeal that uses rolled oats, try making my three-minute old-fashioned oatmeal instead of instant oatmeal packets full of processed sugar. Just add a half-cup each of rolled oats and milk to a large cereal bowl. Stir and cook that mixture for two minutes on high in the microwave. Then add another quarter cup of milk, stir, and microwave for another minute on high. Stir in one to three teaspoons of cinnamon (to taste but be generous), about one to four teaspoons of pure maple syrup or raw honey, and milk if you need to desired consistency. Add fruit (like frozen blueberries run over hot water for a second) and nuts. Enjoy!

Besides good old oatmeal, oats can be added to your diet in many ways. You can add rolled oats to granola, cookies, pancakes, waffles, muffins, bread, and quick bread. You can even use oats with milk, nuts, seeds, and fruit in overnight oatmeal that sits with liquid to soften rather than being cooked. Oats deserve some attention!

RESISTANT STARCH

Cooking and cooling starches creates what is called resistant starches. Resistant starches are similar to fiber. These make pasta, potato, and rice digest slower and decrease how much they make your blood sugar rise—all good things for a leaner body.

BEAT THE RICE AND PASTA RUT

Are you in a rice and pasta rut, searching for ways to eat more whole grains? Start by adding beans, lentils, or edamame to rice or some of the whole grains listed in this section. Also, try these other ideas to beat the carbohydrate rut.

For a fun and nutritious whole grain, try wild, black, or red rice. Brown basmati rice smells like popcorn and has a nice flavor. Use a rice cooker to make it easier on yourself. Make the brown rice taste even better by using no-salt or low-salt stock instead of water to cook it. Or you can branch out beyond rice and pasta altogether. Whole-grain couscous and quinoa are particularly well-suited to substitute for rice, but most whole grains make delicious alternatives as well.

Don't be afraid to prepare whole grains. Most of them can simply be boiled like rice until they're tender. If they start to get too tender, drain the water off; if they're still not tender when there is no water left, just add a little more water. Bigger, chewier grains (called "berries") take about an hour to cook, but they hold their shape well.

Whole grains in a rice cooker

Try using a rice cooker to make whole grains. Add the same amount of water or stock recommended in the stovetop directions. Instant Pots are not as good at cooking whole grains as they are other foods. Adjustments need to be made beyond the regular Instant Pot instructions for successful whole grains. If you want to use an Instant Pot, search the internet for those tricks to make whole grains from bloggers who have figured all the tricks out.

Make whole grains batches

Whole grains keep for about five days, so you can make them in advance and use them throughout the week by mixing them with beans, lentils, and other grains to shake it up. Try them in salads and soups and mixed together in different combinations. Try unfamiliar grains by mixing them in with a grain that you are already accustomed to eating. My favorite whole grain is farro, which cooks in fifteen minutes flat and is both chewy and full of flavor. Kamut, sorghum, spelt, black rice, wild rice, brown basmati rice, amaranth, and a type of young green wheat called freekeh are other delicious ways to free yourself from the rice and pasta rut.

NATALIE'S QUICK QUINOA TIP

Registered dietitian Natalie Stephens makes this quinoa recipe often with extra quinoa she's made during the week. This recipe works with other whole grains you have leftover too. In fact, this recipe would be great with other types of vegetables. It goes with everything, but Natalie often pairs it with salmon or over-easy eggs on super rushed days. Her family loves it!

Sauté chopped onion in about one tablespoon of olive oil for about five minutes. Add a cup of half to quarter-inch broccoli pieces to the pan with another tablespoon of oil. Sauté the broccoli until a little brown. Add a third of a cup of chopped almonds to the pan. When almonds are warm, add two cups of cooked quinoa (or any cooked whole grain). Salt and pepper the quinoa mixture to taste. Yum! Thanks, Natalie!

Great grain baking

If you want to replace all the all-purpose flour in a recipe with sprouted whole wheat, freshly ground whole wheat, or white whole wheat, you need to make a few adjustments. Add two teaspoons of liquid per cup of flour. Next, let the dough rest after mixing for twenty minutes to allow the bran to absorb the extra liquid and avoid any grittiness. This gives you time to clean up the kitchen while your oven preheats.

GRAIN BOWL OPTIONS

Have you noticed how easy it is for some trendy fast-casual restaurants such as Chipotle to throw together a burrito bowl or rice bowl in front of you? You tell them what you like, and they throw it in a bowl. The bowls are delicious and quick. You can make a variation of one at home for a super-easy meal using various whole grains and even leftovers. Use one to two whole grains as a base and then pile on a combination of toppings varying in textures and a balance of salty, sweet, and acidic flavors. Use my simple Grain Bowl Formula to get started.

Grain bowl formula

1. **1 whole grain (or a combination)**

   Cooked whole grains such as sorghum; brown, black, red, or wild rice; quinoa; farro; barley; kasha; wheat berries; rye berries; Kamut; millet; amaranth; and freekeh go in the bowl first. Feel free to mix two or three different whole grains together.

2. **1 to 4 vegetables**

   Top whole grain to cover roughly one-half to three-fourths of the grains. Any raw, steamed, roasted, marinated, sautéed, or pickled vegetable works. For heartier greens such as kale, collards, or cabbage, rub with a little vinegar to tenderize. Roasted vegetables, like Japanese eggplant, beets, and broccoli, are magnificent in these bowls! As are lentils, chickpeas, beans, and dried peas of any sort, and pickled vegetables (such as kimchi or pickled cauliflower) or marinated vegetables such as artichoke hearts to add lots of flavor to the bowl.

3. **1 to 2 types of protein**

   Place next to the vegetable(s). Choose a creamy cheese such as ricotta or fresh mozzarella, poultry, pork, beef, lamb or fish (leftover or freshly cooked), eggs (over-easy, fried, soft boiled, poached, or scrambled), chickpeas, beans, lentils, dried peas, edamame, tofu, tempeh, or seitan. Smoked meats and strongly seasoned meats go a long way to add flavor to the whole bowl.

4. 0 to 2 fruits

Avocado offers delicious creaminess, but other fruits add a bit of sweetness to balance the flavors. Pineapple, mango, pomegranate, papaya, chopped apples, pitted and sliced Medjool dates, and dried cranberries or cherries are a few ideas, but don't limit yourself.

5. 0 to 3 sauce /dressing

Olive oil, infused or regular balsamic vinegar, other types of vinegar, pesto, infused olive oil, vinaigrette, tahini sauce, sesame oil, peanut sauce, hoisin sauce, teriyaki sauce, fish sauce, tamari, soy sauce, tamarin, liquid amino, salsa, chutney, anchovy paste, chili-garlic sauce, gochujang (Korean chili sauce), harissa, Sriracha, miso, hummus, baba ghanoush, plain yogurt, fresh lime, or lemon can be used to add flavor and moisture to your bowl. My cashew-based dressing is amazing on grain bowls! Go to www.FoodsWithJudes.com and search for Guilt-Free Creamy Dressing.

6. 2 to 3 toppings

Add 1 crunchy and 1 to 2 for flavor. Preserved lemon; pickled vegetables such as pickled peppers or radishes; toasted seaweed; nutritional yeast; fresh herbs; cheese such as feta, goat, or Halloumi; chutney; and Kalamata olives add flavor, while sesame seeds, sunflower seeds, nuts, and pomegranate add crunch.

FIX 10

CHANGE GRAINS

Eat between three to six servings of whole grains per day instead of refined carbohydrates.

Make the switch and replace your refined grains like white, refined bread and buns, white rice, and white flour pasta with whole grains like whole-grain bread, black or brown rice, ancient grains, and whole-grain pasta.

Whole grain benefits

The fiber and phytochemicals in these whole grains help to release glucose slowly into the bloodstream, providing a steady supply of energy to the body and brain. The fiber in whole foods is especially effective in slowing down the absorption of sugar and keeping insulin from spiking. Those who eat more fiber gain less weight over time.

Eating whole grains every day gives you many health benefits. Beyond helping to control blood sugar, these undefiled grains lower the risk of type 2 diabetes, heart disease, high blood pressure, and cancer. Whole grains help control weight but eating them instead of refined, processed grains cuts down on the amount of body fat around your middle.

Whole grains are an important food for the health-promoting microbes in the gut. They increase good bacteria in the large intestine for better digestion, greater absorption of nutrients, and a stronger immune system. Refined carbs, on the other hand, feed the bad bacteria and contribute to an imbalance of microbiota in your gut that can negatively affect your weight, ability to fight off sickness, risk of disease, and even your responses to the world around you. Switch your grains to whole grains; your gut depends on them.

What is a serving?

One serving of whole-grains is about half a cup or one slice of whole-grain bread. Thinly sliced whole-grain bread is two slices per serving, and popcorn is three cups per serving.

Keep track of the number of servings of whole-grain foods you eat each day if you want to lose or maintain inches, but don't get too worried about measuring your food. Once you know how much a half a cup looks like, just estimate rather than measuring your servings out and making yourself crazy.

When you eat refined, processed grains, such as white bread or white rice, rather than a whole grain on occasion, the refined grain should be counted as part of the three to six grains you eat each day if you want to lose or maintain inches. That said, please give yourself a few refined grains every week and enjoy them without guilt! You can be flexible in your life while still meeting your wellness goals. Get rid of the all-or-nothing mentality, and give yourself some flexibility so that you can boost your health and keep inches off for life.

How many servings

Eat at least three servings of whole grains per day but no more than six servings if you want to maintain or lose inches. These whole grains are filled with thousands of phytonutrients, fiber, vitamins, and minerals that truly make a difference to your energy level, thinking power, weight, body composition, inflammation levels, disease state, and gut health. Don't be tempted to skip them! Taking whole grains out of your diet completely is analogous to taking vegetables out of your diet. But because all carbs are demonized, the true power of whole grains is often lost or at least misunderstood.

Whole grains for fuel

More than six servings may be too much if you want to lose inches. For those engaging in moderate exercise less than an hour per day and who are of average to small build, this three to six serving range should be about right.

If you're doing intense exercise bouts, you may need more than six servings of whole grains. If you are a large-framed person and in athletic training, you will almost certainly need more than six servings of whole grains. If you aren't in training but exercise strenuously, you still may need to increase your servings of whole grains on those high-movement days. The whole grains protect your muscles from being broken down for energy so the protein you eat can be used to repair and build more muscle.

Sprouted grains and fermented grains

Sprouted grains catch the sprouts during the germination process, breaking down some of the carbohydrates and chemicals like lectins and phytate to make the grains easier to digest and more nutritious than regular whole grains. Fermented grains, made by friendly microbes feeding on the carbohydrate of the grains, also increase nutrient content and make whole grains easier to digest.

The good carbs!

Whole grains are the good guys! They help you decrease body fat, absorb fewer calories than refined carbs, fight disease, and keep your mind alert, while foods made with refined carbs increase your body fat and contribute to illness and grogginess. Plus, whole grains are more satisfying than processed grains, so you can count your whole grains and not worry about getting hungry!

Action plan

- ✓ Eat three to six whole-grain half-cup (or one slice of bread) servings per day. Eat more than six servings of whole grains if you are adding intense exercise, add more whole grain servings for long-lasting fuel.
- ✓ Include ancient grains as a part of your daily whole-grain intake.
- ✓ Count your whole grain servings correctly. One serving of cooked whole grains is about half a cup or one piece of bread. Thinly sliced whole-grain bread is two slices per serving and popcorn is three cups per serving.
- ✓ Make oats a part of your regular food line up.
- ✓ Prepare and eat one new whole grain every month.
- ✓ Try mixing unfamiliar whole grains with familiar grains such as rice.
- ✓ Cook extra whole grains and use them throughout the week. They stay good in the refrigerator for five days.
- ✓ Check the ingredient list on a food product for whole grains. Avoid products that include refined grains in their products.
- ✓ Eat oats in some form as part of your whole servings when possible.
- ✓ Try sprouted whole grain bread for a healthier option. Check the label for more details of how it's made and the grains used to make it.
- ✓ Eat sourdough bread made with whole grains and a long fermentation time as part of your whole grain servings. Those times when you eat white bread, at least choose a sourdough bread variety.
- ✓ Track your progress on the *12 Fixes to Healthy* app.

FIX 11
MUNCH MINDFULLY

There always seems to be a holiday, vacation, or party peeking around the corner ready to sabotage your wellness goals! But you don't need to give up the fun just to keep your health and weight on track.

I have some simple techniques you can use to enjoy the fun times in your life without sabotaging your wellness goals. These tricks actually help you eat less while giving you more enjoyment, all without fighting your willpower! So no matter what you're eating, you can enjoy your food more!

I should warn you, though—don't dismiss these tips because they seem too simple! These techniques really do work!

Fix 11 is to chew small bites of food slowly and completely, only eat when you're hungry, and stop eating just before you get too full.

CHEW AND SWALLOW

We tend to chew our food a few times and then swallow, already anticipating our next bite. Does that sound familiar? You probably wouldn't guess that chewing small bites of food slowly could help you enjoy your food more and cause you to eat less with more satisfaction. But that's exactly what happens!

Step one

Step one is to eat small bites. That might mean to eat one piece of popcorn or nut at a time or half the bite you typically take now. That's not so hard, right?

Step two

Step two is to pay attention to when you actually swallow your bite of food. Do you swallow it before your bite is completely chewed? You should chew your bite until it's essentially liquid or nothing solid is left. That is your clue to swallow.

Most people swallow long before their bite is completely chewed. So start noticing when exactly you're swallowing your food.

Step three

Step three is to make sure that you're not taking another bite or putting more food into your mouth before you have completely chewed and swallowed the previous bite.

> **CHEW TEST**
>
> See how long it takes for you to chew a baby carrot, some almonds, an apple slice with peanut butter, or a gummy bear when you don't swallow it until there are absolutely no chunks. It will surprise you how long it takes to chew up just one gummy bear. That is more time to enjoy it and more time for your body to send the signal that you've had enough food.

Chew more, eat less

Extra chewing tends to slow down the pace of your eating. Research studies report that people who eat slowly tend to eat less. You might assume that the participants took larger bites, but in one study the bites were all the same size and they ate the same food.

When these same participants chewed each bite forty times as instructed, all the participants ate less regardless of their weight. When they chewed their food a little more than normal, their levels of gut hormones related to hunger and satiety also improved.[1]

Another study looked at the effects of eating speed on the amount of food eaten. Participants who ate slowly tended to eat less and were less hungry sixty minutes after starting the meal.[2]

> **MIND ILLUSIONS**
>
> Chewing your food for a longer period of time tricks your mind into thinking that you're eating more than you're actually eating. If you think you're eating more, you actually end up eating less. More chewing makes you more satisfied with the meal even though you ate less.

Chew more, weigh less

Chewing more may also help you keep inches off. Another team of researchers followed 529 participants for eight years, finding that those who ate slowly gained less weight than those eating quickly. The participant's BMI, exercise and drinking habits had less effect on their weight loss than the speed at which they ate.

Eating quickly in this study was linked to a 35 percent increase in risk for metabolic syndrome, a group of health issues including excess abdominal fat, elevated blood sugar levels, high blood pressure, and lab values that point to heart disease.[3]

Chew more, burn more

Extra chewing may even result in burning an extra ten calories during digestion of a three-hundred calorie meal. That doesn't seem like enough calories to blink an eye at, but if you do the math, you could burn about two thousand extra calories each month, according to researchers.[4]

> **CHEWING BURNS CALORIES**
>
> In a study reported by the *American Journal of Nutrition,* participants who chewed each bite of their food forty times ate 12 percent fewer calories than did those who chewed their food only fifteen times per bite. That might not seem like a lot, but 10 to 12 percent fewer calories consumed over a few years may translate into a pants size.[5]

DIGESTION BOOST

Digestion actually starts in the mouth. Chewing helps to break down food and blend it with an enzyme in saliva that breaks down carbohydrates. Skipping this step forces your stomach to work even harder to do the job. Even smoothies should be chewed a little!

Chewing your food completely triggers the full production of stomach acid and pancreatic enzymes in the digestive tract to help you digest your food better further down the track. Chewing fully results in an adequate amount of blood flowing to the stomach and gut to aid in your digestion.

Chewing well also breaks down the food smaller, so when you swallow the food, it mixes more thoroughly with your stomach acid. Stomach acid breaks down food into smaller, more digestible molecules as well as keeping pathogens in your food under control.

Slow down and take a moment before you gobble up your food. Inhaling food too quickly doesn't allow you time to mentally process that you have eaten. Slowing down helps you to make eating more of a conscious process and helps you to eat more mindfully.

HUNGER SIGNALS

Learning to get in touch with your hunger signals is essential for getting you to a healthy weight—and keeping you there. All of us are capable of telling when we are hungry and full. You just need to relearn how to listen to your body, the way you did when you were a baby and toddler.

By the time we are about three years old, we start to ignore our bodies' hunger signals. Americans are conditioned by our culture to eat regardless of whether we are actually hungry. Most of us grew up with TV commercials promoting processed foods, super-sized portions from restaurants, and well-meaning parents telling us to clear our plates.

All those things are external cues. One study found that people who rely on external cues to tell them when to stop eating are heavier than are those who rely on internal cues.[6]

THE APPLE TEST

When you are craving a snack, try the "apple test" to see if you are truly hungry. If you are tempted to grab a snack, question yourself by asking, "Am I hungry enough to eat an apple?" If an apple sounds good, then you are likely physically hungry. Honor your hunger by asking, "Am I hungry for something hot, cold, sweet, savory, soft, or crunchy?" Maybe think of a pear or an orange if you don't prefer apples.

Linda Allen RDN/LN, CDE, ACE

SPEND LESS TIME REGRETTING WHAT YOU EAT AND MORE TIME LEARNING FROM IT.

NATALIE STEPHENS, RD/LDN

Researchers interviewed 133 participants from Paris and 145 participants from Chicago. They asked them, "How do you know when you are through eating dinner?" and received two distinctly different responses. The Parisians said they knew when they were done with a meal "when I feel full or when the food no longer tastes good." Chicagoans answered that they were done with a meal "when everyone else (at the table) is done or when my plate is empty."

The Parisian group of participants was leaner than the American group. That's not a coincidence!

What does your body say?

Look at your eating patterns. Do you tend to focus on a computer or TV screen while eating rather than focusing on when your body is full? Do you usually eat a snack every day at two in the afternoon—even when you're not hungry? When you really think about it, you may find that you eat when you "feel" like it or based on mindless habits instead of when you're truly hungry.

How you respond to external eating cues can have a big impact on your weight. If you want to avoid overeating, pay more attention to what your body is telling you rather than to what is left on your plate!

MINDFUL EATING

Mindfulness is a way to "savor the moment" and get more pleasure out of the same amount of food or less of unhealthy food. It reconnects you to your internal cues for eating and makes it easier to eat only out of hunger and to stop with satiety.

MINDLESS EATING VS MINDFUL EATING

Eating beyond your body's full signals	**VS**	Listening to your internal cues and stopping when full
Eating based on emotions (i.e. sad bored angry)	**VS**	Eating when you are physically hungry
Eating foods that are emotionally comforting	**VS**	Eating nutritious foods
Eating while doing other activities	**VS**	Giving attention to the tastes of your food
Eating alone at unplanned places and times	**VS**	Eating with others at set places and times.

Let's try it

Choose a favorite food, like chocolate. You can use two different types of chocolate to get a feel for comparing and noticing the different flavors—because that is what mindfulness is all about. I also really like to do this with one cold strawberry and one warm/room temperature strawberry.

1. Pull out your inner child and pretend like you are seeing it for the first time—maybe think about when you introduce new foods to a baby and how they respond. Make a note of the color, shape, appearance, and even how it feels in your hand.

2. Hold it to your nose and smell it. Do any memories or feelings come to mind?

3. Place and hold in your mouth—just in the middle of your tongue. What textures and tastes do you notice?

4. Now chew it a few times, then stop. Can you taste it more, less? How much pressure does it put against your teeth, your cheeks?

5. Chew more and move it around your mouth. Is the flavor getting stronger? Notice if you are salivating more and if it coats your tongue. Is there a spot on your tongue where it tastes the best?

6. Swallow and repeat.

7. On a scale of 1-10, how much did you enjoy that? How did it compare to your usual way of eating? Did you notice compulsive feelings, anxious feelings, feelings of gratitude or calmness? How does your stomach feel afterward?

> ## EATING FOR THE DOUBLE DOPAMINE EFFECT
>
> Taste isn't the only source of pleasure we get from our food. Your body releases a big dose of dopamine twenty minutes after eating nourishing foods too. Mindful eating allows you to tap into both sources of pleasure. Eating well becomes instinctive and enjoyable without rules!
>
> Natalie Stephens, MS, RD, dietingdifferently.com

This is where the Hunger Scale comes in. How full are you? Heavy, satisfied, energized? How does the rest of your body feel?

> ## WHOLE-BODY EXERCISE
>
> Here is your first exercise. Take an inventory of what is happening in your whole body.
>
> **Head:** Is your head tired, stressed, sluggish, or energized? Are you focused or everywhere?
>
> **Nose:** Is your nose stuffy, runny, or clear?
>
> **Mouth/lips:** Are your lips dry or comfortable?
>
> **Shoulders:** Are your shoulders up by your ears? Knotted or relaxed?
>
> **Chest:** How are your heart rate and breathing? Are you tense, anxious, relaxed?
>
> **Belly:** Are you bloated, hungry, full, constipated? Are your clothes comfortable?
>
> **Arms/legs:** Do your arms feel stiff or flexible, strong or tired?
>
> **Feet:** Do you feel tired, sore, comfy?

HUNGER SCALE

1 Ravenous, weak, dizzy

2 Very hungry, cranky, low energy, stomach growling

3 Mildly hungry; you want food, but you can wait a little longer

4 Feeling a little hungry

5 Satisfied, not hungry or full

6 Slightly full

7 A little uncomfortable

8 Feeling stuffed and uncomfortable

9 Very uncomfortable, stomach in pain

10 Very full; you feel sick

Hunger Scale

Use the Hunger Scale to help identify how your body is directing your eating. Identify your hunger level using the code.

When your hunger is at a 3 or a 4, eat something. Don't wait until you're feeling at a 1 or a 2. If you let yourself get too hungry, you risk overeating. Then, stop eating when you reach 5 or 6 on the scale. Identify what "satisfied" or "slightly full" feels like for you.

When you train your brain to eat based on your body signals, you prevent weight gain and digestive issues. Plus, it feels great to be in control of your body instead of letting external signals control your eating—and consequently, your health.

NEW WAY OF THINKING

Getting rid of your all-or-nothing mentality is essential to making lasting, meaningful lifestyle changes. This plan is not just about following the 12 Fixes in this book. It's about shifting your thinking and self-talk toward praising yourself as you gradually make positive changes, rather than focusing on past choices that weren't health promoting.

No matter what food choices you make, you can always apply this Fix to Munch Mindfully without feeling stressed! In fact, you will enjoy your food more by paying attention to your food and to how your body feels. Who doesn't want to enjoy their food more?

Perfection is overrated

In the case of these 12 Fixes, perfection is not only overrated but unnecessary. Following these fixes most of the time for the rest of your life gives you a similar outcome as if you followed them 100 percent of the time for the rest of your life. Furthermore, following these fixes most of the time throughout your life offers a much better outcome than following these fixes 100 percent of the time for a period of time before going back to your old ways.

In reality, perfectionism often throws you off the plan altogether! It brings you down and prevents you from feeling positive about the good choices you make. Focusing on being perfect hurts you way more than it helps you. Eating isn't set up for perfection, so you're bound to make a "mistake," but you don't have to think of it as a mistake. Eating nourishes and fuels your body, and it connects us with people. We all need to eat, and thinking of food as negative is harmful for you and self-sabotaging. Feeling guilt leads to ongoing shame and stress, which leads to hopelessness and self-loathing. You end up binge eating with no motivation to follow

NOTHING TASTES BETTER WITH A SIDE OF GUILT! CHOOSE THOUGHTFULLY AND ENJOY YOUR FOOD.

RUTH RANKS MED, RD, CSOWM

any of the 12 Fixes. So in the end, aiming for perfection is much worse than following the 12 Fixes most of the time and allowing yourself to enjoy your less-healthy indulgences here and there.

- strict
- deprivation
- hunger and cravings
- misery
- guilt
- boring
- social isolation
- quick fix
- being thin
- good and bad foods

Change the messages around eating healthy

What associations do you have with the word "diet"? Here are a few that come to mind for me. Do any of these associations resonate with you?

All these associations are not only incorrect but severely limit your ability to enjoy healthy food and notice all the great benefits of doing so! Let's take all the negative associations and put a spin on it so you can really approach food as a source of nourishment instead of a source of punishment!

How does that feel?

Doesn't that sound and feel so much better? Aren't you more inspired and motivated when you reframe healthy eating? Re-examine all the negative beliefs you have around healthy eating and see if there is a better way to look at it. It will take time to replace these well-ingrained negative thoughts, but be prepared to correct those beliefs and replace them with empowering and more accurate messages.

ASSOCIATIONS WITH THE WORD "DIET" AND A HEALTHIER APPROACH

Strict (100% or nothing!)	**VS**	80:20 or MOST of the time for the rest of my life
Deprivation (I can't have what I want)	**VS**	What do I really want? What is going to get me what I want?
Hunger and cravings (for food)	**VS**	Nourish & care for myself so I can get what I really crave out of life
Misery (I'm giving up what I want)	**VS**	I'm trading something I want now for something I want more later
Guilt (I made a mistake, I am bad)	**VS**	I have an opportunity to learn something, which gets me one step closer to my goal
Boring (It's a plan you follow and that's it)	**VS**	It's a journey of exploring enjoyable ways to incorporate healthy food and activity
Social isolation (I can't eat what others eat)	**VS**	I'm an example that can inspire and invite others to live healthier too
Quick fix	**VS**	Way of being
Being thin (I need to change)	**VS**	Being strong & the best version of me I can be
Good and bad foods	**VS**	Foods all have trade-offs; which do I want?

Your adventure

Eating certain foods is no different than turning left or right. It isn't good or bad; it just takes you to different places. Sometimes you need to stick to a certain path to get what you want, and other times you can take a detour or come up with your own way entirely. It isn't any different with eating. For certain results (like if you have kidney failure or if you have food intolerances) you will need to follow a more precise way of eating and living to feel your best. There are also times where it is good to get off "track" and just enjoy a treat with some friends. And for most of us, there is no one perfect way to eat, so you get to create your own way from the guidelines laid out here. And that is an adventure in and of itself!

Body love

Food is literal and symbolic nourishment. That is the reason we use it to show care and concern. To feed something or someone is to love it. The way you see yourself and your body will impact your ability to feed yourself well. If you don't believe you are worth being loved, it's hard to believe that you are worth caring for yourself and feeding yourself well.

How many times have you heard someone say that thin people can eat what they want and heavy people need to be on a diet? Or that thin people can wear what they want but heavy people need to cover up? Those ideas are being challenged, but they are still widely-held opinions. Hidden in all those messages is the idea that heavy people should be limited in what they eat and do. That seems more like a punishment, yet nourishing your body should not be a punishment. Those two ideas really can't coexist.

Come up with ten things that are good about you and that are worth growing and nurturing. Don't hesitate to reach out to your support system for perspective, because we sometimes struggle to see the good in ourselves.

The way you see yourself and your body will impact your ability to feed yourself well. The opposite is also true. The way you feed yourself will impact how you see yourself. If you feed yourself well, you are symbolically telling yourself that you deserve to be nourished. Use food as an opportunity to tell yourself that you deserve to be nourished and that you deserve to enjoy your food, your body, and your life.

DON'T LET CRAVINGS CONTROL YOU!

Eating 30 to 40 grams of protein at breakfast, staying well hydrated, and getting enough sleep will give you a winning advantage over your cravings by reducing impulsivity in your brain and by balancing appetite hormones.

When cravings come, try giving yourself permission to eat it if you wait twenty minutes first, and consider whether you still want it. This helps in three ways. First, telling yourself "don't eat it" is really only understood by your brain as "eat it." Telling yourself not to do it just sets up a power struggle that drains your self-control.

Second, you will learn that the craving usually goes away. You will learn to trust that it won't be long until you don't feel the need to eat it anyway. Just drink a big glass of water or some herbal tea and eat a handful of almonds, and then do something that takes your attention. You may not even think of it again.

Third, if you still want that food after twenty minutes, or if you just can't wait for twenty minutes, it gives you a chance to practice mindful eating and self-care. Go get that food, then sit down without distractions and savor every aroma, texture, and flavor!

Then decide if you really liked it as much as you thought you would. If you didn't like it as much as you hoped, you either need to be pickier about your indulgent foods or maybe your body needs something else. Maybe what you really need is sleep, a good laugh, or a relaxing walk. Or you may need better skills for coping with stress. Giving in to cravings and then listening to your body gives you an opportunity to sort this out.

Some of the cravings that we experience are simply a conditioned response to certain places, times of day (or year!), people, or feelings. If it is conditioned, that means you can recondition yourself! Understand your cravings better with this exercise the next time you have one.

CRAVING QUESTIONS EXERCISE

1. What is it you want?
2. When did you first notice you wanted it?
3. Who were you with and where were you?
4. What time of day was it?
5. What were you feeling at that time (hunger, stress, happiness)?

These questions should help you get clearer on which associations trigger the craving. Just identifying the triggers will give you a better perspective and more control over your response.

When you identify the trigger, you can strategize new responses. For example, if you start craving a cookie as soon as you leave work, pack some peppermint or lavender lotions for stimulating aromatherapy. Or pull out a five-minute meditation exercise on your phone. If it is a certain person who triggers the craving response, maybe talk to them about finding alternative foods or activities you can enjoy together. If it is a certain time of day, set an alarm on your phone to remind you to just be aware and to try something new. The more enjoyable the new alternative action, the sooner the old craving will go away.

FIX 11

MUNCH MINDFULLY

Chew small bites of food slowly and completely, only eat when you're hungry, and stop eating just before you get full.

To maintain a healthy weight throughout your life, you need to regulate the amount of food you eat based on how hungry you are—not on the amount of food within your reach.

Chewing slowly gives your brain more time to realize that you're full. You've probably heard that before, but I bet you don't realize how powerful the impact is. Try to chew your food until there's nothing left to chew anymore. You will be surprised at the large chunks of food that slip down your throat even when you try to eat slowly.

Chewing your food for a longer time tricks your mind into thinking you're eating more than you're actually eating. If you think you're eating more, you actually end up eating less. Ironically, even though you consume less food, research shows that more chewing makes you more satisfied with the meal. Speed-eating beats the body's "full" signals to the punch so you eat more food before hormones can let us know you are full. Chewing buys time for your brain to tune in and for digestive juices to secrete, making it easier for your body to digest your food, leading to the absorption of more nutrients and better digestion.

Practice mindful eating, and take your time chewing as you take in your food. Literally "savor the moment" and get more pleasure from less food by letting yourself enjoy your food!

Hunger scale and mindfully munching

Don't revert to the old "eat everything on your plate" adage. Stop eating when your body starts to feel full. Use this Hunger Scale on page 216 to judge how hungry or full you feel.

Use the Hunger Scale to help identify how your body can direct your eating. Don't wait until you're starving at a 1 or a 2 on this scale, and stop eating before you are too full at a 7 - 10. Stop eating when you are at about a 5 or 6 on the scale. Identify what "satisfied" or "slightly full" feels like for you.

Stop using external cues—such as the time of day and the amount of food on your plate—to decide when and how much you eat. Instead, listen to your body's internal cues to determine when you're actually hungry and satisfied. Keep in mind, you can eat the remaining food later when you are actually hungry. It tastes better when you are hungrier, anyway! Using your internal cues to dictate the amount of food you eat, coupled with eating the right types of foods, can help you live a far healthier life.

Action Plan

- ✓ Chew your food mindfully and slowly, for longer amounts of time. Avoid doing anything while eating except for socializing. Mindful eating means paying attention to each morsel, not looking at your screen.
- ✓ Take small bites. Eat one piece at a time.
- ✓ Chew until your food is liquified before you swallow.
- ✓ Make sure you don't take a bite before the previous bite has been completely chewed and swallowed.
- ✓ Put your utensils down between every bite. Sip water between bites, but don't wash your food down with your drink.
- ✓ Only eat when you're hungry—that's a 3 or 4 on the Hunger Scale.
- ✓ When you sit down to a meal, first stop and think about how hungry you are. If you're not very hungry, make a conscious effort to eat less than usual.
- ✓ Relax before you start eating and enjoy every morsel. Don't rush.
- ✓ Pause a quarter of the way through your meal and analyze your hunger. If you're still hungry, keep eating, but stop if you're not. You don't have to clean your plate no matter what your parents taught you.
- ✓ Stop eating when you reach a 5 or 6 on the Hunger Scale. Get in touch with what "satisfied" or "slightly full" feels like for you.
- ✓ Don't hesitate to decide to finish your meal as a future snack or meal later when you are hungrier. The food will taste better when you are hungrier anyway.
- ✓ If you have difficulty breaking the habit of eating despite feeling full, identify what other reason you are eating for. Try to come up with non-food solutions to address the reason for eating when full.
- ✓ Track your progress on the *12 Fixes to Healthy* app.

FIX 12
SLEEP SOUNDLY

Do you struggle to get enough quality sleep? One in five Americans sleeps less than six hours on a typical weeknight.[1] Yet sleep is so critical to your health, weight, and cognitive function!

While the brain is in deep sleep, it does some heavy-duty housekeeping. The waste products that build up in the spaces between brain cells while you're awake get cleared out during sleep. Beta-amyloid is a sticky substance that forms plaques between brain cells and impedes communication between neurons. A buildup is linked to damaged brain function and Alzheimer's disease. Sleep plays a critical role in cleaning beta-amyloid out of the brain before it forms plaques.

Fix 12 is to sleep seven to nine hours every night to help keep destructive plaques from building in the brain.

Sleep also helps you function, feel good, and make healthy choices the following day. Sleep, movement, and diet are all interlaced. Movement affects how you sleep, and sleep affects how you move. Sleep impacts what you eat, and what you eat affects your sleep. A lot is riding on a good night's sleep!

MOVE MORE TO SLEEP BETTER

Movement and sleep are intertwined, enhancing each other. Consistent movement can be key to a better night's sleep. Regardless of age and demographics, movement has been consistently associated with better sleep, more than any other behavior.[2]

The reverse is also true. We've all experienced the lack of energy and motivation that comes with a sleepless night. It's hard to be productive, much less to exert energy for a workout, without adequate sleep. On the other hand, you can have more energy to move and to enjoy doing it when you are well rested. No amount of exercise will make up for a poor night's sleep!

While a good workout can help you to be more alert, to energize you for the day, and to up your metabolism, nighttime workouts don't appear to disrupt your sleep. One study reported that vigorous exercise even thirty-five minutes prior to bed did not upset the participants' sleep patterns.[3]

KEEP IT COOL!

Your core body temperature drops during sleep, allowing your body to move into more restorative REM and slow-wave sleep. Studies confirm that when you crank the heat, you wake up more often and you sleep less deeply. Keep the thermostat at around 68 degrees Fahrenheit for the best sleep results.[4]

SNOOZE TO LOSE

Have you ever noticed how difficult it is to resist nutrient-empty junk food when you're tired? Or how much easier it is to make positive food choices when you're rested?

Sleep empowers us to eat better, burn fat, and build muscle. When you get enough sleep and movement, you are energized and optimistic. You're happier, sharper, and more likely to make your health a priority and find greater success in your work and relationships.

Sleep can be your medicine or your poison. The first eleven fixes introduced as part of the *12 Fixes to Healthy* plan are all about making small changes for a big return. Sleep is no different. Somehow, it has become a badge of honor in our society to get by on less sleep. Yet the bottom line of your health and success is directly related to the amount of sleep you get. Getting enough sleep ensures that you have sufficient energy to live to the fullest.

When you have energy from a good night's sleep, you tend to make wiser choices with your movement and nutrition. The length and quality of your sleep have a direct impact on how you experience life. When you have enough sleep, you improve your memory, spur creativity, sharpen your attention, and improve your motivation and performance. You also improve your long-term health with enough sleep.

Chronic inflammation,[5] insulin resistance,[6] and obesity[7] all decrease with sleep.[8] Sleep directly affects your muscle and body-fat levels. If you snooze, you lose!

> **SLEEP DEPRIVATION AND CAFFEINE SURPRISE!**
>
> According to Dr. Michael A. Grandner, director of the Sleep and Health Research Program at the University of Arizona College of Medicine, sleep deprivation dulls your thinking skills. Caffeine doesn't help you think better; it just helps you make poor decisions faster.[9]

Brain function

Sleep plays a critical role in memory, both before and after learning. Both the quantity and quality of sleep affect learning and memory. Without sleep, you cannot focus your attention sufficiently to learn or unify new information enough to recall it. It's diffictul to learn and retain information.[10] With sufficient sleep, you can consolidate memories and strengthen them. The emotional components of memory are also strengthened during sleep, which may benefit the creative process.[11]

Lack of sleep can result in inebriation-type symptoms. Sleep-deprived individuals find it more difficult to concentrate, assess situations, and make decisions. Lack of adequate sleep even affects mood, motivation, judgment, and our perception of events. Without enough sleep, life seems harder. Lack of sleep contributes to depression, irritability, moodiness, and anxiety.

> **TIRED OR DRUNK?**
>
> According to the National Highway Traffic Safety Administration, tired drivers' performance accounts for the highest number of fatal crashes in some years. Sleeplessness affects reaction times and decision-making. In fact, the effect of too little sleep one night on a driver is similar to that of drinking alcohol.[12]

Muscle and body fat

Skipping out on adequate sleep can lower your muscle mass[13] and can result in extra visceral fat around your middle.[14] During sleep—particularly REM sleep—your body repairs and builds muscle while also breaking down fat for energy. Without enough quality sleep, you're unable to build as much of that valuable muscle or to burn as much fat. If you are up and not sleeping, you are more likely to eat late into the night, compounding the problem and leading to more nutritionless calories along with less opportunity for your body-fat stores to be used.[15]

Studies have shown a strong correlation between length of sleep and amount of body fat. Those who slept less than eight hours per night had a higher level of body fat. One study in particular, out of the University of Chicago, compared body-fat percentage with sleep amount in the same participants on a restricted diet. Participants lost 55 percent less fat and 60 percent more lean tissue when they slept five and a half hours per night.[16]

Hormones

Several studies have investigated why we tend to eat more when tired. Lack of sleep often causes hormone imbalances that affect us negatively. These studies have found a strong relationship between the activity of the hunger hormone (ghrelin) and the number of hours of sleep. Furthermore, without enough sleep, the fullness hormone (leptin) levels drop, while hunger hormone levels increase.[17]

Given the hormone imbalance, it's not surprising that we eat more when skimping on sleep and significantly increase the number of calories we consume. A study analyzing the results of eleven studies found that sleep-deprived individuals ate almost four hundred extra calories per day compared with people who got enough sleep. Researchers reported the sleep-deprived individuals ate higher fat, lower protein foods and were prone to food cravings.[18]

Your stress hormones are also impacted by your sleep habits. Deep sleep neutralizes stress hormones, but lack of sleep triggers the release of stress hormones such as adrenaline and cortisol, increasing the risk of heart disease, belly fat, and muscle breakdown.[19] Because stress hormones increase fluid retention, you can actually gain two pounds in body fluids from just one night of poor sleep.

Caving to cravings

Increased hunger hormone levels don't just make you hungrier. They also make you crave junk food. Researchers found that the desire for high-carbohydrate, high-calorie foods increased by a whopping 45 percent in individuals experiencing high levels of the hormone ghrelin.[20]

Sleep studies using brain imaging found that the reward-seeking portion of the brain is stimulated by fewer than seven hours of sleep per night, increasing the desire for refined-carbohydrate foods that spike blood sugars and invite fat-promoting insulin.[21] One of these studies found that completely sleep-deprived participants showed diminished activity in the critical thinking, decision-making part of the brain as well as greater activity in pleasure-seeking areas.[22] It appears that the less sleep you get, the more likely you are to act on your food cravings and eat more processed foods.

NIGHTTIME EATING

Your body shifts from eating mode into sleep mode during the evening. Keep your evening meal light, and skip those nighttime snacks to improve your sleep. Researchers found that people, especially women, who eat during the evening have lower quality sleep.[23]

Eating high-fat foods in the evening is particularly disruptive to your sleep. Studies found that eating dietary fat in the evening disrupted sleep more than merely eating food in the evening did. Both women and men who consumed higher amounts of fat during the evening spent more time in bed without sleeping compared to those who ate lower-fat foods or had no food at all during the evening.[24]

Illness

Not only does lack of sleep compromises your immune system, making sickness harder to fight, but you may not recover as quickly either. Too little sleep reduces cytokine production, which reduces the immune response to infection.[25] Researchers from the University of California, San Francisco found that those who get six hours or less of sleep during the night are four times more likely to catch a cold when exposed to a virus compared to those who slept more than seven hours.[26]

Sleep is central to your ability to keep fat weight off, to function well, and to be productive. For most of us, it's a matter of making it a priority. All you need to do is sleep for seven or eight hours. If you can plan ahead to go to bed at an earlier time most nights, you can reap all the many life-enhancing benefits that a full night of sleep has to offer.

The simple yet powerful act of sleep can either enhance or sabotage your health and your success. In fact, all aspects of our lives are affected, so let's aim for a good night's sleep.

FOOD TO SNOOZE

The relationship between sleep and diet is reciprocal. Your diet appears to influence quality of sleep too.

A study published in 2016 in the *Journal of Clinical Sleep Medicine* found that diet can influence sleep. This study reported that eating less fiber, more saturated fat, and more sugar is associated with a lighter, less restorative night's sleep with more disruptions. The primary researcher from Columbia University Medical Center, Marie-Pierre St-Onge, PhD, noted her surprise that a single day of lower fiber, higher dietary fat intake could influence sleep.

This study reported that more fiber intake predicted more time spent in the stage of deep sleep. More saturated fat was associated with less deep, slow-wave slumber. Greater sugar intake also was associated with an increased number of arousals from sleep.

This randomized crossover study had participants spend nine hours of sleep in a lab in bed. They snoozed for seven hours and thirty-five minutes on average per night after three days of controlled eating and after one day of eating whatever they wanted.

The study also found that participants fell asleep twelve minutes faster after eating fixed meals lower in saturated fat and higher in protein. Other studies have also found fiber, sugar, and protein intake to affect sleep.[27]

SLEEPY FRUIT

Certain foods have been specifically identified with better sleep. Tart cherries and kiwifruits contain properties that enhance sleep. High concentrations of melatonin and high antioxidant capacity may be responsible for tart cherries' positive effects on sleep. Kiwifruit may promote sleep because of its unusually high serotonin content and high antioxidant capacity. Kiwifruits are also full of folate. Low levels of folate are linked to insomnia and restless leg syndrome.[28]

Fiber

The National Health and Nutrition Examination Surveys (NHANES) is a series of studies designed to assess the nutritional status and health of people in the US. In 2007–2008, over twelve thousand participants with the best sleep habits had the highest fiber intake, whereas those who slept five to six hours or more than nine hours had lower fiber intake. Participants who slept less than five hours per night had the lowest fiber intake.[29]

Sugar

The NHANES 2007–2008 analysis found that those who slept five to six hours per night had higher sugar intake than the best sleepers.[30] Consistent with this finding, the NHANES 2005–2010 analysis reported that higher total sugar intake was associated with those who slept six hours or less per night. However, it's unclear if or how sugar impacts sleep, since those sleeping less than five hours had lower sugar intake than normal sleepers in the NHANES 2007–2008.[31]

> ### DARK CHOCOLATE
>
> While a little dark chocolate during the day may be beneficial to your taste buds and your health, nighttime dark chocolate isn't good for your sleep. Not only does dark chocolate contain sugar and fat, but it also contains the stimulant theobromine, which can pump your heart rate and disrupt your sleep. Shucks!

Protein

Protein intake has also been linked to sleep duration and sleep quality. In NHANES 2007–2008, participants who slept less than seven hours per night had a lower total protein diet.[32] A study evaluating 410 young adult women reported less than six hours of sleep for those participants consuming a lower protein, higher carbohydrate diet compared to normal-sleeping participants.[33]

A cross-sectional analysis of over 4,000 non-shift workers in Japan found that those consuming a low protein intake of less than 16 percent of calories from protein had difficulty getting to sleep and experienced poor quality sleep.[34] It appears we do need enough protein to sleep well. More research is needed, but an amino acid (tryptophan) is a precursor to the sleep hormone melatonin and may be partly responsible for the effects of protein on sleep.[35]

Mediterranean diet

The Mediterranean diet is a way of eating based on the traditional diets of countries bordering the Mediterranean Sea. Like the food-centered fixes in this book, the Mediterranean diet is a diet high in vegetables, fruits, whole grains, beans, lentils, nuts and seeds, olive oil, seafood, Greek plain yogurt, and small amounts of high-quality cheeses.

Repeated research shows a correlation between the Mediterranean diet and health-promoting sleep. Individuals that followed a Mediterranean diet experienced less insomnia and achieved healthier sleep durations in several studies. These studies show only associations (rather than a cause and effect link) between sleep and the Mediterranean diet, but there's some evidence to suggest that the high fiber, low saturated fat of the Mediterranean diet may be responsible for these observations.[36]

The anti-inflammatory and antioxidant effects of the Mediterranean diet may promote healthy sleep. For instance, olives and grapes are an important part of the Mediterranean diet and are rich in melatonin that assists your circadian cycle.[37] A Mediterranean pattern of eating is lower in processed foods and higher in foods closer to their whole-food form, which is associated with a less stressed life with more organization. This may have an important impact on better sleep patterns.

Caffeine

As much as you may want to drink caffeine during the day to wake up, it may affect your sleep. People consuming caffeine during the day sleep less with less quality and have more daytime sleepiness than do their non-caffeine-consuming counterparts.[38]

Individuals vary widely in their sensitivity and tolerance to caffeine, but sensitivity to caffeine may not be so obvious. Caffeine may not decrease your amount of sleep, but may subtly affect your quality of sleep without you even knowing it. Caffeine consumption appears to interrupt your crucial slow-wave deep sleep.

Slow-wave sleep is the brief, critical period of deep sleep needed to reset and resynchronize. During this period, your brain removes memories from short-term memory and stores them away, similar to cleaning up your desktop at the end of the day. This deep sleep is when the brain's house-cleaning service does its deep clean to get rid of the sticky stuff that forms harmful plaque, impeding communication between the neurons.

Given individual variations in caffeine tolerance, sleep patterns of individuals more sensitive to caffeine may be disrupted, even with small amounts of caffeine consumed earlier in the day. Caffeine can negatively affect your deep sleep, so go easy on it even if you don't think of yourself as caffeine-sensitive!

The American Academy of Pediatrics suggests that children under the age of twelve years not drink any caffeine-containing drinks. For teens, caffeine intake should be no more than a hundred milligrams per day, the equivalent to an 8-ounce cup of home brewed coffee or two 12-ounce sodas.

That individual tolerance variance to caffeine holds true for children and teenagers, so don't be misled by the influx of caffeine-packed drinks available now. Some teens experience sleep problems with even a can or two of soda, much less an energy drink.

ALCOHOL AND NICOTINE

While alcohol consumption before sleep makes it easier to fall asleep quickly, alcohol is associated with poor-quality, disrupted sleep. Nicotine showed the greatest disturbance to sleep in a study evaluating caffeine, alcohol, and nicotine on sleep.[39] So go easy on these beverages and use this study as another reason to not smoke!

OPERATION SHUT DOWN

Maintaining healthy eating and exercise patterns can truly help you be a better sleeper. But turning your brain off can be very difficult. The truth is that you've inadvertently programmed your brain to keep going while you're in bed. Consequently, you can be really exhausted, but as soon as you hit the pillow, your brain turns to GO mode.

How many of you do other tasks in bed besides sleep? Most people do some of the following while in bed: television/movies, podcasts, reading, YouTube videos, eating, working, arguing, worrying, talking, planning, replaying events, texting, and rehearsing. That is a lot of activity for a bed that should be reserved for sleep and sex. We've trained our brains to do so many things in bed besides sleep!

You need to reprogram your brain to know that the bed is meant for sleeping and not all the other activities that you might be tempted to do in bed. You can get so comfortable watching a movie, for instance, that you don't hear your body's tired signals. Imagine standing up on your bed watching a show. You would hear your body's exhaustion signals loud and clear. Even watching in the living room on the sofa is more likely to allow your body to tell you that it's time to stop for the night and head to your sleep haven. Plus, the couch is not your bed. Your bed should be reserved for sleeping.

To reprogram your brain to know that the bed is only for sleeping, make a concentrated effort to do other activities somewhere else! This change in habit can take some time to adopt, but if you stick with it, your brain will more often respond to the feeling of your bed by setting the body to sleep mode, even at an early hour.[40]

WORRY LESS

When you don't give yourself time to think and process during the day, guess when you do it? Yes, at night, when the lights are out and it's quiet! Deirdre Conroy, clinical director of the Behavioral Sleep Medicine Clinic at the University of Michigan Health System recommends allocating time during the day to make a "worry" list. Come up with possible solutions to the pressing matters on the list during the day so by the time you go to bed, you can put your mind and body to rest.

FIX 12

SLEEP SOUNDLY

Sleep seven to nine hours every night.

Let your brain sleep and clear out the plaque that builds up in the spaces between your brain cells while awake. Otherwise, this sticky stuff builds up over time and interferes with communication between your neurons.

Sleep can make or break your day and your future health. Your rest affects your willingness to exercise, what food and drink choices you make, how much fat you burn, and even how much muscle you build. The reverse is also true. Your food and workout choices affect the quality and quantity of your sleep. Move more and get to bed earlier!

Get to bed

Set a time to go to bed, and stick to it. The hardest part of this fix is developing the discipline to turn off the TV or the computer screen at the end of the night. Blue-light screens from smartphones and computers right before bed are especially problematic for sleep. This kind of light sends the same signal to your body as sunlight, so your entire sleep cycle ends up delayed into the next morning. There is a function on some smartphones that can dim the phone automatically and use non-blue light on a schedule.

Leaving the phone in another room and getting a good old-fashioned alarm clock isn't a bad idea, either. Kids and teenagers especially should be leaving their phones in rooms other than where they sleep at night.

Avoid nighttime eating

Avoid eating at night, especially foods that are high in fat, refined carbohydrates, and sugar. These foods raise your blood sugar level and keep you from sleeping well. Additionally, stop eating early in the evening so that you are burning fat longer during the night. Eating plenty of protein during the day boosts your levels of sleep-friendly hormones such as melatonin and serotonin and make sure to move more to improve your sleep.

Caffeine, alcohol, and nicotine

Individuals vary widely in their sensitivity and tolerance to caffeine, but caffeine sensitivity may not be so obvious. Caffeine may not decrease your amount of sleep, but may subtly affect your quality of sleep without you even knowing it. Caffeine consumption can interrupt slow-wave deep sleep needed for your brain to reset, resynchronize, and clean out harmful toxins. So even though you want to drink caffeine to wake you up during the day, it may just be making you more tired the next day and in need of more caffeine.

While alcohol consumption before sleep makes it easier to fall asleep quickly, alcohol is associated with poor-quality, disrupted sleep. Nicotine showed the greatest disturbance to sleep in the research evaluating caffeine, alcohol, and nicotine on sleep. So go easy on these beverages and don't smoke to obtain a more rejuvenating night's sleep!

Not tired?

If you're not tired near bedtime, engage in some relaxing activities that put you in the mood to sleep. Read a book, listen to relaxing music, or meditate. Keep lights low or off if possible. Wi-Fi LED bulbs, which last many years, are getting cheaper. There are dimmable ones and even dimmable color bulbs that can adjust away from blue light. They can be changed with your phone with a device like Alexa or on an automatic schedule. Plus, they don't use very much energy.

Try smelling lavender essential oil; it can help you relax in preparation for sleep. Sleep in darkness, and keep the temperature in your room around 68°F when possible. If you have pets, try to keep them off the bed or out of the room so that they don't disturb you.

Reprogram your brain

Make your bed a place to sleep and not for activities like television, movies, radio/podcasts, reading, eating, working, arguing, worrying, talking, thinking, planning, replaying events, texting, and rehearsing. Changing your habits takes time, but you can do it. You taught your brain that bed was the time and place for some of those activities, and you can reprogram your brain too!

Action Plan

- ✓ Sleep seven to nine hours during the night as often as possible. Make it a priority! You may need to give yourself a bedtime.
- ✓ Stay away from blue-light screens at least two to three hours before going to bed, or change the screen to white or yellow light.
- ✓ Turn off phones, computers, and TVs during the night.
- ✓ Keep lights dim or off during the hour prior to bedtime, and block noise using earplugs.
- ✓ Turn down the thermostat when possible, as a cool room promotes deeper sleep.
- ✓ Get completely ready for bed at least one hour before bedtime and read, write, etc.
- ✓ Stick to a relaxing ritual prior to sleep.
- ✓ Resist the temptation to squeeze one more task into the end of your day.
- ✓ Avoid eating during the evening hours after a light dinner. If you do need to eat something, avoid sugar and refined carbohydrates.
- ✓ Eat a diet full of vegetables, fruits, whole grains, beans, lentils, nuts and seeds, olive oil, seafood, Greek plain yogurt, and small amounts of high-quality cheeses.
- ✓ Infuse lavender essential oil in your bedroom or place some on a tissue and tuck the tissue into your pillowcase prior to bedtime.
- ✓ Keep pets off the bed during the night.
- ✓ Track your progress on the *12 Fixes to Healthy* app.

SUCCESS

A few years ago, I married a wonderful man who definitely wasn't a daily eater of fresh, high-quality, delicious food.

When we met, he was a business executive who didn't take much time for nutritious food and exercise. He ate out a lot, at both fast-food and sit-down restaurants filled with hamburgers, french fries, and steaks. At home, he ate his favorite go-tos: cereal every morning and plastic-wrapped American cheese slices toasted on white bread for lunch. He often ended his day with his biggest meal, quickly cleaning his plate even if he was already satisfied.

After we met and he learned my 12 Fixes, he began to move more and made small, simple changes to what, how, and when he ate. He didn't diet; he adopted the 12 Fixes little by little, not all at once. He made small changes that led to better habits without much effort. None of the dieting all-or-nothing mentality, and no deprivation! I didn't shame him for occasionally eating an old favorite meal or snack, instead, I just encouraged him to eat more deliciously prepared vegetables and to try new foods. He realized that healthier, unprocessed foods kept him more satisfied throughout the day, and he began to gravitate toward healthier options that I kept available. Each fix made him feel more satisfied and energetic.

After just nine months of following some of the fixes some of the time, he had very noticeable improvements in his health. For his whole adult life, he had high cholesterol (LDL) levels, but his LDL levels were now within the normal range.

While his weight on the scale remained about the same, his waist size steadily diminished. He was amazed as he continued to move his belt notches down until, finally, he was forced to get a new belt. He went down a full pants size in eight months, without the pain of dieting, having made many little easy changes.

The food we ate was tasty, and we didn't spend much time in the kitchen making it. I wasn't his personal chef either! Eventually, he could often be found in the kitchen making my quick recipes and searching for healthier options at restaurants.

At first, my husband was surprised by how much he enjoyed healthier foods and that he didn't miss his old favorites. He realized the ease of changing his ways most of the time, but not 100 percent of the time. What was most impressive to him was how much better he felt. He hadn't realized eating better and moving more could feel so good!

My husband now understands that he does NOT have to be on a rigid diet—or any diet for that matter—to get leaner, healthier and stronger. He became more aware that the number on the scale isn't a good indicator of his progress compared with measuring the inches around his waist.

He's thrilled with how he feels and looks. These are just the immediate benefits of eating right. Because these changes are sustainable he will continue to reap the benefits; he will save money and misery with fewer medications and fewer doctor visits. He lives stronger and maintains more independence as he ages. He is less likely to have to battle cancer, kidney disease, or liver disease. He will be living a better-quality life with fewer health issues, and I'll be right there with him enjoying the journey!

YOU'VE GOT THIS!

It's time to transform your life for the better.

Using the *12 Fixes to Healthy* plan laid out in this book, making changes will be so much easier than you realize and more enjoyable than you could imagine. You'll feel, think, and look better as you gain muscle, regardless of your present weight and figure. Who doesn't want a better quality of life with more strength and fewer health issues?

Just start with one fix at a time!

Doable

Sometimes our lives are complicated. I've tried to simplify these powerful fixes to help you integrate them into your life with ease. All twelve of these fixes are important and can improve your life. However, adding even just one of these fixes can benefit your health and weight if you try to do it most of the time. Doing them most of the time is so much more doable and pleasurable than trying to follow them 100 percent of the time. This wiggle room is just enough leeway to keep you enjoying this plan for life without guilt.

Enjoyable

These fixes will add pleasure to your life rather than pain. Don't think of this plan as a diet! Focus on all the new rewarding choices you add. There's no need to dwell on the times that you make a less healthy choice. This is not an all-or-nothing endeavor.

Don't stress about falling off the wagon for this plan, because there is no wagon! It's not a strict plan that keeps you fighting your will-power. In fact, it's designed for you to create new, rewarding, pleasurable habits and routines that will keep you satisfied without any guilt if you occasionally eat or drink something unhealthy. Pat yourself on the back for all the great decisions you make if you're trying to keep good habits 80-90 percent of the time. It's easy when you add one fix at a time!

Flexible

No matter what you eat, you can enjoy your food and eat less with greater satisfaction. In fact, Fix 8, *Shift Surroundings* and Fix 11, *Munch Mindfully*, contain strategies in regard to how to serve and eat your food no matter what you're eating. So even if you are indulging, there are some painless tricks that help you eat less with more pleasure!

Fix 5, *Eat Early*, provides strategies around when you eat and drink, so again, it's not always about what you eat. Fix 6, *Move More*, and Fix 12, *Sleep Soundly*, are also non-food fixes that boost your health and improve your weight immensely. All three of these non-food fixes help you make better food and beverage choices, without giving anything up.

Regarding the five fixes concerning what you eat, you will be surprised how good and satisfying those foods can be, with surprising flexibility. No matter what diet restriction you have, these 12 Fixes can be woven into your life.

After experiencing all 12 of these Fixes, you will feel amazing, look wonderful, think smarter, and be happier overall. You'll love the lifestyle and will want to follow it every day. You'll feel like you can conquer the world.

12 FIXES TO HEALTHY

I've put these 12 Fixes in a specific order to give you the biggest benefit for the least amount of work. However, no matter what order you add them in, they will be powerful to you. The critical key for success is spending some time focusing on integrating the fixes into your life, one fix at a time.

Concentrate on one fix for a week to a month, then focus your attention on the next fix while continuing the first one. If you give each fix some time to work it into your routines, you can continue doing each one without the attention they originally needed. That frees up your ability to focus on the next fix while still following the ones you added already. Don't forget that you shouldn't go for perfection, just about 80 percent of the time! No guilt is necessary for this *12 Fixes to Healthy* plan.

READY, SET, GO!

Give yourself at least twelve weeks, but, ideally, twelve months, to add the fixes to your life. Then continue working on the fixes that are trickier for you, focusing on one at a time. You need time to focus on adopting each fix by creating new, rewarding routines. Below are the instructions to maximize this *12 Fixes to Healthy* plan.

1. Begin by measuring your waist with a measuring tape. Take other body measurements too, if you wish. Write down these measurements, and remeasure exactly the same way every month or even every three months if that is more comfortable for you. If you have access to a body composition test, feel free to find out your body fat and lean tissue body percentages to begin, and then retest every three months along with tracking your waist measurements.

   Don't be fooled by the number on a scale. This isn't a good indicator of the body fat you are losing. A gain in your compact muscle (which you want) can mask the fat weight you lose, making it seem like you didn't lose as much body fat as you actually lost.

2. Add and focus on just one of these fixes for a specified time period—at least one week up to one month. You can add them in any order you want; however I have arranged them in an advantageous order for most people. As long as you add on (not replace) one fix each week to a month, this *12 Fixes to Healthy* plan will work well for you.

3. Identify some specific ways to make the "focus fix" fit your lifestyle before you add it (see the action plans for each fix introduced throughout each of the previous chapters).

4. Use the *12 Fixes to Healthy* app to track the fixes you add. Continue to follow and track the ones you've already added.

5. After you have had a chance to experience and track all twelve of these fixes, continue following all twelve fixes. Because it takes about thirty days to form a new habit, concentrate on ingraining the fixes you struggle with into your

routine, one fix each month until all twelve fixes become an integral part of your life.

6. Make it a point to tell someone about the fix you are focusing on. Encourage your friends and coworkers to follow it along with you and share your ideas with them. Let your friends, family, and coworkers be a means of accountability, and encourage each other to keep at it. Help one another concentrate on the positive changes you're making. Don't forget to give each other a pat on the back to celebrate each small success!

7. Use the *12 Fixes to Healthy Workbook*, and take advantage of the videos, recipes, and meal plans integrating all 12 Fixes.

Make each fix your norm

You don't need to be perfect with the fix you're focused on, but do try to be extremely mindful of it. Remember, the idea is to make a habit and a new routine out of each fix you add. This will make it easier for you to turn each fix into the norm for the rest of your life rather than being a temporary exception to the rule.

That norm doesn't mean following a fix 100 percent of the time. Following these fixes, even 75 to 80 percent of the time for the rest of your life will make a huge difference in your body composition, productivity, and your overall health.

Too easy trap

Don't get stuck thinking that any of these fixes or ways to adapt these fixes are too insignificant or too easy to really make a difference. It's a trap! Doing something, however small, is better than doing nothing. Small, sustainable changes over a long time make a bigger difference than lots of changes (hard or easy) over a short period of time.

Perfection not required!

Don't be discouraged if you are not perfect. You don't need to be perfect for this plan to succeed! Celebrate your successes instead of dwelling on failures. In fact, there are no failures in this plan. Track the things you do right, not the things you do wrong, with the *12 Fixes to Healthy* app. Before you know it, guilt will vanish, and health and success will become a part of your everyday living.

12 Fixes to Healthy promotes a leaner, healthful, and productive life and is about feeling empowered by the health-promoting choices you make every day—even the little ones. Every fix you make, large or small, is going to have a positive impact on your well-being in the long run.

FIX 1 **SWAP SUGAR**
Replace sugar, refined white flour, and processed foods with whole foods.

FIX 2 **PLAN PROTEIN**
Eat a total of 75 to 150 grams of protein spread throughout the day, with at least 30 grams of protein for breakfast.

FIX 3 **BOOST GUT**
Eat a fermented food, like yogurt, every day.

FIX 4 **PUSH PRODUCE**
Make fruits and vegetables 50 percent of your meals and snacks.

FIX 5 **EAT EARLY**
Frontload your eating, and avoid eating in the evening.

FIX 6 **MOVE MORE**
Move your body more.

FIX 7 **WATER WELL**
Drink water throughout the day and evening.

FIX 8 **SHIFT SURROUNDINGS**
Rearrange, resize, plan, and search your environment.

FIX 9 **SWITCH FAT**
Replace less healthy fats with healthier fats.

FIX 10 **CHANGE GRAINS**
Eat between three and six servings of whole grains per day instead of refined carbohydrates.

FIX 11 **MUNCH MINDFULLY**
Chew small bites of food slowly and completely, only eat when you're hungry, and stop eating when you're satisfied, not too full.

FIX 12 **SLEEP SOUNDLY**
Sleep seven to nine hours every night.

REFERENCES

Introduction

1. Dariush M, Hao T, Rimm EB, Willett WC, Hu FB. Changes in Diet and Lifestyle and Long-Term Weight Gain in Women and Men. *N Engl J Med.* 2011;364(25):2392-404. doi:10.1056/nejmoa1014296.

Chapter 1 Swap Sugar

1. Boyer J, Liu RH. Apple phytochemicals and their health benefits. *Nutr J.* 2004;3(1). doi:10.1186/1475-2891-3-5.
2. Liu RH. Health benefits of fruit and vegetables are from additive and synergistic combinations of phytochemicals. *Am J Clin Nutr.* 2003;78(3):517S-20S. doi:10.1093/ajcn/78.3.517S.
3. Bhupathiraju SN, Tucker KL, Katherine L. Greater variety in fruit and vegetable intake is associated with lower inflammation in Puerto Rican adults. *Am J Clin Nutr.* 2010;93(1):37-46. doi:10.3945/ajcn.2010.29913.
4. Hung HC, Joshipura K, Willett W. RESPONSE: Fruit and Vegetable Intake and Risk of Major Chronic Disease. *J Natl Cancer Inst.* 2004;96:1577-584. doi:10.1093/jnci/dji107.
5. Bhupathiraju SN, Wedick NM, Pan A, Manson JE, Rexrode KM, Willett WC, Rimm EB, Hu FB. Quantity and variety in fruit and vegetable intake and risk of coronary heart disease. *Am J Clin Nutr.* 2013;98(6):1514-523. doi:10.3945/ajcn.113.066381.
6. Hall et al. Ultra-Processed Diets Cause Excess Calorie Intake and Weight Gain: An Inpatient Randomized Controlled Trial of Ad Libitum Food Intake. *Cell Metab.* 2019;30(2):67–77. doi:10.1016/j.cmet.2019.05.008.
7. Quirk SE, Williams LJ, O'Neil A, Pasco JA, Jacka FN, Housden S, Berk M, Brennan SL. The Association between Diet Quality, Dietary Patterns and Depression in Adults: a Systematic Review. *BMC Psychiatry.* 2013;13(1). doi:10.1186/1471-244X-13-175.
8. Jacka FN, O'Neil A, Opie R, Itsiopoulos C, Cotton S, Mohebbi M, Castle D, et al. A Randomised Controlled Trial of Dietary Improvement for Adults with Major Depression (the 'SMILES' Trial). *BMC Med.* 2017;15(1). doi:10.1186/s12916-017-0791-y.
9. Francis HM, Stevenson RJ, Chambers JR, Gupta D, Lim CK. A Brief Diet Intervention Can Reduce Symptoms of Depression in Young Adults – A Randomised Controlled Trial. *Public Library of Science.* Accessed November 11, 2019.
10. Ahmed SH, Guillem K, Vandaele Y. Sugar addiction: pushing the drug-sugar analogy to the limit. *Curr Opin Clin Nutr Metab Care.* 2013;16(4):434-39. doi:10.1097/mco.0b013e328361c8b8.
11. Volkow ND, Wang GJ, Fowler JS, Tomasi D, Baler R. Food and Drug Reward: Overlapping Circuits in Human Obesity and Addiction. *Curr Top Behav Neurosci.* 20122:1-24. doi:10.1007/7854_2011_169.

    Lenoir M, Serre F, Cantin L, Ahmed SH. Intense Sweetness Surpasses Cocaine Reward. *PLoS ONE.* 2007;2(8). doi:10.1371/journal.pone.0000698.

12. Hu FB. Resolved: there is sufficient scientific evidence that decreasing sugar-sweetened beverage consumption will reduce the prevalence of obesity and obesity-related diseases. *Obes Rev.* 2013;14(8):606-19. doi:10.1111/obr.12040.
13. Wirfält E, Drake I, Wallström P. What do review papers conclude about food and dietary patterns? *Food Nutr Res.* 2013:57. doi:10.3402/fnr.v57i0.20523.

    Volkow ND, Wang GJ, Tomasi D, Baler RD. The Addictive Dimensionality of Obesity. *Biol Psychiatry.* 2013;73(9):811-18. doi:10.1016/j.biopsych.2012.12.020.

14. Benton D, Ruffin MP, Lassel T, Nabb S, Messaoudi M, Vinoy S, Desor D, Lang V. The delivery rate of dietary carbohydrates affects cognitive performance in both rats and humans. *Psychopharmacology (Berl).* 2003;166(1):86-90. doi:10.1007/s00213-002-1334-5.
15. Diano S, Farr SA, Benoit SC, McNay EC, da Silva I, Horvath B, Gaskin FS et al. Ghrelin controls hippocampal spine synapse density and memory performance. *Nat Neurosci.* 2006;9(3):381-388. doi:10.1038/nn1656.

    Benton D, Ruffin MP, Lassel T, Nabb S, Messaoudi M, Vinoy S, Desor D, Lang V. The delivery rate of dietary carbohydrates affects cognitive performance in both rats and humans. *Psychopharmacology (Berl).* 2003;166(1):86-90. doi:10.1007/s00213-002-1334-5.

16. Shan B, Cai YZ, Sun M, Corke H. Antioxidant Capacity of 26 Spice Extracts and Characterization of Their Phenolic Constituents. *J Agric Food Chem.* 2005;53(20):7749-7759. doi:10.1021/jf051513y.

    Pham AQ, Kourlas H, Pham DQ. Cinnamon supplementation in patients with type 2 diabetes mellitus. *Pharmacotherapy*. 2007;27:595–599. doi:10.1592/phco.27.4.595.

Cinnamon. National Center for Complementary and Integrative Health. https://nccih.nih.gov/health/cinnamon. Published November 12, 2019. Accessed November 21, 2019.

Kirkham S, Akilen R, Sharma S, Tsiami A. The potential of cinnamon to reduce blood glucose levels in patients with type 2 diabetes and insulin resistance. *Diabetes Obes Metab*. 2009;11:1100–1113. doi:10.1111/j.1463-1326.2009.01094.x.

17. Mohapatra DP, Thakur V, Brar SK. Antibacterial Efficacy of Raw and Processed Honey. *Biotechnology Research International*. 2011;2011:1-6. doi:10.4061/2011/917505.

18. Atkinson FS, Foster-Powell K, Brand-Miller JC. International tables of glycemic index and glycemic load values: 2008. *Diabetes Care*. 2008;31(12):2281-2283. doi:10.2337/dc08-1239.

19. Al-Waili NS. Natural Honey Lowers Plasma Glucose, C-Reactive Protein, Homocysteine, and Blood Lipids in Healthy, Diabetic, and Hyperlipidemic Subjects: Comparison with Dextrose and Sucrose. *J Med Food*. 2004;7(1):100-07. doi:10.1089/109662004322984789.

20. Madjd A, Taylor MA, Delavari A, Malekzadeh R, Macdonald IA, Farshchi HR. Beneficial Effects of Replacing Diet Beverages with Water on Type 2 Diabetic Obese Women Following a Hypo-energetic Diet: A Randomized, 24-week Clinical Trial. *Diabetes Obes Metab*. 2017;19(1):125-132. doi:10.1111/dom.12793.

21. Ruiz-Ojeda FJ et al. Effects of Sweeteners on the Gut Microbiota: A Review of Experimental Studies and Clinical Trials. *Adv Nutr*. 2019;10: S31-S48. doi:10.1093/advances/nmy037.

22. Harvard Health Publishing. Abundance of fructose not good for the liver, heart. Harvard Health. www.health.harvard.edu/heart-health/abundance-of-fructose-not-good-for-the-liver-heart. Accessed April 27, 2017.

Schaefer EJ et al. Dietary fructose and glucose differentially affect lipid and glucose homeostasis. *J Nutr*. 2009;139(6):1257S-1262S. doi:10.3945/jn.108.098186.

Pollock NK et al. Greater fructose consumption is associated with cardiometabolic risk markers and visceral adiposity in adolescents. *J Nutr*. 2012;142(2):251-7. doi:10.3945/jn.111.150219.

23. Ventura EE et al. Sugar Content of Popular Sweetened Beverages Based on Objective Laboratory Analysis: Focus on Fructose Content. *Obesity*. 2010;19(4):868–874. doi:10.1038/oby.2010.255.

24. Center for Food Safety and Applied Nutrition. Labeling & Nutrition - Changes to the Nutrition Facts Label. U.S. Food and Drug Administration. www.fda.gov/food/food-labeling-nutrition/changes-nutrition-facts-label. Accessed April 27, 2017.

25. Jonnalagadda SS, Harnack L, Hai Liu R, Mckeown N, Seal C, Liu S, Fahey GC. Putting the Whole Grain Puzzle Together: Health Benefits Associated with Whole Grains--Summary of American Society for Nutrition 2010 Satellite Symposium. *J Nutr*. 2011;141(5). doi:10.3945/jn.110.132944.

26. Scarlata K. What's Behind Wheat Sensitivities? *Today's Dietitian*. 2017;19(5):28. www.todaysdietitian.com/newarchives/0517p28.shtml.

Specter M. Against the Grain. The New Yorker. October 27, 2014. www.newyorker.com/magazine/2014/11/03/grain. Accessed March 23, 2017.

27. Burkhart A. The Microbiome and Celiac Disease: A Bacterial Connection. Amy Burkhart, MD, RD & SMS. theceliacmd.com/microbiome-celiac-disease-bacterial-connection. Published October 28, 2019.

Santa Cruz, J. The Gut Microbiome's Link to Celiac Disease. *Today's Dietitian*. 2019;21(5):24. www.todaysdietitian.com/newarchives/0519p24.shtml.

28. Nanda R, Shu LH, Thomas JR. A FODMAP Diet Update: Craze or Credible? *Prac Gastroenterol*. 2012;:37.

Gibson PR, Shepherd SJ. Evidence-based dietary management of functional gastrointestinal symptoms: The FODMAP approach. *J Gastroenterol Hepatol*. 2010;25(2):252-258. doi:10.1111/j.1440-1746.2009.06149.x.

Chapter 2 Plan Protein

1. Paddon-Jones D, Westman E, Mattes RD, Wolfe RR, Astrup A, Westerterp-Plantenga M. Protein, weight management, and satiety. *Am J Clin Nutr*. 2008;87(5):1558S-561S. doi:10.1093/ajcn/87.5.1558S.

2. Paddon-Jones D, Rasmussen BB. Dietary protein recommendations and the prevention of sarcopenia. *Curr Opin Clin Nutr Metab Care*. 2009;12(1):86-90. doi:10.1097/MCO.0b013e32831cef8b.

3. Pratley R, Nicklas B, Rubin M, Miller J, Smith A, Smith M, Hurley B, Goldberg A. Strength training increases resting metabolic rate and norepinephrine levels in healthy 50-to 65-yr-old men. *J Appl Physiol (1985)*. 1994;76(1):133-137. doi:10.1152/jappl.1994.76.1.133.

Campbell WW, Crim MC, Young VR, Evans WJ. Increased energy requirements and changes in body composition with resistance training in older adults. *Am J Clin Nutr*. 1994;60,(2):167-175. doi:10.1093/ajcn/60.2.167.

4. Westcott WL. Why The Confusion on Muscle and Metabolism? www.yumpu.com/en/document/view/46619084/why-the-confusion-on-muscle-and-metabolism-wayne-l-westcott-

5. Layman DK, Evans EM, Erickson D, Seyler J, Weber J, Bagshaw D, Griel A, Psota T, Kris-Etherton P. A moderate-protein diet produces sustained weight loss and long-term changes in body composition and blood lipids in obese adults. *J Nutr*. 2009;139(3):514-21. doi:10.3945/jn.108.099440.

6. Layman DK. Dietary Guidelines should reflect new understandings about adult protein needs. *Nutr Metab (Lond)*. 2009;6(1):12. doi:10.1186/1743-7075-6-12.

7. Martin WF, Armstrong LE, Rodriguez NR. Dietary protein intake and renal function. *Nutr Metab (Lond)*. 2005;2(1):25. doi:10.1186/1743-7075-2-25.

8. Antonio J, Ellerbroek A, Silver T, et al. The effects of a high protein diet on indices of health and body composition – a crossover trial in resistance-trained men. *J Int Soc Sports Nutr*. 2016;13:3 doi:10.1186/s12970-016-0114-2.

9. Vander Wal JS, Gupta A, Khosla P, Dhurandhar NV. Egg breakfast enhances weight loss. *Int J Obes*. 2008;32(10):1545-1551. doi:10.1038/ijo.2008.130.

10. Leidy HJ, Ortinau LC, Douglas SM, Hoertel HA. Beneficial effects of a higher-protein breakfast on the appetitive, hormonal, and neural signals controlling energy intake regulation in overweight/obese, "breakfast-skipping," late-adolescent girls. *Am J Clin Nutr*. 2013;97(4):677-688. doi:10.3945/ajcn.112.053116.

11. Leidy HJ, Racki EM. The addition of a protein-rich breakfast and its effects on acute appetite control and food intake in 'breakfast-skipping' adolescents. *Int J Obes (Lond)*. 2010;34(7):1125-1133. doi:10.1038/ijo.2010.3.

    Leidy HJ, Bossingham MJ, Mattes RD, Campbell WW. Increased dietary protein consumed at breakfast leads to an initial and sustained feeling of fullness during energy restriction compared to other meal times. *Br J Nutr*. 2009;101(06):798-803. doi:10.1017/s0007114508051532.

    Leidy HJ, Ortinau LC, Douglas SM, Hoertel HA. Beneficial effects of a higher-protein breakfast on the appetitive, hormonal, and neural signals controlling energy intake regulation in overweight/obese, "breakfast-skipping," late-adolescent girls. *Am J Clin Nutr*. 2013;97(4):677-688. doi:10.3945/ajcn.112.053116.

12. Zeisel SH, Da Costa K. Choline: an essential nutrient for public health. *Nutr Rev*. 2009;67(11):615-623. doi:10.1111/j.1753-4887.2009.00246.x.

13. Galioto R, Spitznagel MB. The Effects of Breakfast and Breakfast Composition on Cognition in Adults. *Adv Nutr*. 2016;7(3):576S-89S. doi:10.3945/an.115.010231.

14. Layman DK. Dietary Guidelines should reflect new understandings about adult protein needs. *Nutr Metab (Lond)*. 2009;6(1):12. doi:10.1186/1743-7075-6-12.

15. Institute of Medicine Food and Nutrition Board. *Dietary Reference Intakes: Energy, Carbohydrates, Fiber, Fat, Fatty Acids, Cholesterol, Protein, and Amino Acids*. Washington, DC: National Academies Press; 2002. doi:10.17226/10490. Accessed March 2, 2020.

16. Layman DK. Dietary Guidelines should reflect new understandings about adult protein needs. *Nutr Metab (Lond)*. 2009;6(1):12. doi:10.1186/1743-7075-6-12.

17. Arentson-Lantz E, Clairmont S, Paddon-Jones D, Tremblay A, Elango R. Protein: A Nutrient in Focus. *Appl Physiol Nutr Metab*. 2015;40(8):755–61. doi:10.1139/apnm-2014-0530.

18. Layman DK. Dietary Guidelines should reflect new understandings about adult protein needs. *Nutr Metab (Lond)*. 2009;6(1):12. doi:10.1186/1743-7075-6-12.

19. Thomas DT, Erdman KA, Burke LM. Position of the academy of nutrition and dietetics, dietitians of canada, and the american college of sports medicine: Nutrition and athletic performance. *J Acad Nutr Diet*. 2016;116(3):501-528. doi:10.1016/j.jand.2015.12.006.

20. Mcguire S. U.S. Department of Agriculture and U.S. Department of Health and Human Services, Dietary Guidelines for Americans, 2010. 7th Edition, Washington, DC: U.S. Government Printing Office, January 2011. *Adv Nutr*. 2010;2(3):293-94. doi:10.3945/an.111.000430.

21. Simopoulos AP, Salem N Jr. n-3 fatty acids in eggs from range-fed Greek chickens. *N Engl J Med*. 1989;321(20):1412. doi:10.1056/NEJM198911163212013.

22. Gorman, RM. Are Cage-Free Eggs Really Better?. *EatingWell*. www.eatingwell.com/article/289956/are-cage-free-eggs-really-better. Published January 2017. Accessed January 25, 2020.

23. A Consumer's Guide to Food Labels and Animal Welfare. Animal Welfare Institute. awionline.org/content/consumers-guide-food-labels-and-animal-welfare. Accessed January 25, 2020.

    "Free Range" and "Pasture Raised" officially defined by HFAC for Certified Humane® label. Certified Humane. certifiedhumane.org/free-range-and-pasture-raised-officially-defined-by-hfac-for-certified-humane-label/. Published October 9, 2014. Accessed December 17, 2019.

24. Penn State. Research shows eggs from pastured chickens may be more nutritious. Penn State News. news.psu.edu/story/166143/2010/07/20/research-shows-eggs-pastured-chickens-may-be-more-nutritious. Published February 21, 2020. Accessed February 22, 2020.

25. Karsten HD, Patterson PH, Stout R, Crews G. Vitamins A, E and Fatty Acid Composition of the Eggs of Caged Hens and Pastured Hens: Renewable Agriculture and Food Systems. *Renewable Agriculture and Food Systems*. 2010; 25(1):45-54. doi:10.1017/S1742170509990214.

26. Kühn J, Schutkowski A, Kluge H, Hirche F, Stangl GI. Free-Range Farming: A Natural Alternative to Produce Vitamin D-Enriched Eggs. *Nutrition*. 2014;30(4):481-4. doi:10.1016/j.nut.2013.10.002.

27. Simopoulos AP, Salem N Jr. n-3 fatty acids in eggs from range-fed Greek chickens. *N Engl J Med*. 1989;321(20):1412. doi:10.1056/NEJM198911163212013.

28. U.S. Food and Drug Administration. Advice about Eating Fish. www.fda.gov/food/consumers/advice-about-eating-fish. Published July 2, 2019. Accessed January 25, 2020.

29. Ralston NV, Raymond NJ. Dietary selenium's protective effects against methylmercury toxicity. *Toxicology*. 2010;278(1). doi:10.1016/j.tox.2010.06.004.

    Ralston N. Selenium: The Secret That Will Change Public Perception of Seafood. Talk presented at: Academy of Nutrition and Dietetics 2016 Food and Nutrition Conference & Expo; October 16, 2016; Boston, MA.

    Alehagen U, Johansson P, Björnstedt M, Rosén A, Post C, Aaseth J. Relatively high mortality risk in elderly Swedish subjects with low selenium status. *Eur J Clin Nutr*. 2016;70(1):91-96. doi:10.1038/ejcn.2015.92.

    Hoffmann PR, Berry MJ. The influence of selenium on immune responses. *Mol Nutr Food Res*. 2008:52(11):1273-1280. doi:10.1002/mnfr.200700330.

30. Bonjour, JP. Dietary Protein: An Essential Nutrient For Bone Health. *J Am Coll Nutr*. 2005;24(6):526S-536S. doi:10.1080/07315724.2005.10719501.

31. O'Connor LM, Lentjes MAH, Luben RN, Khaw KT, Wareham NJ, Forouhi NG. Dietary dairy product intake and incident type 2 diabetes: a prospective study using dietary data from a 7-day food diary. *Diabetologia.* 2014;57,5:909-17. doi:10.1007/s00125-014-3176-1.

    Dennett C. The Truth About Dairy Fats. *Today's Dietitian.* 2016;18(10):26. www.todaysdietitian.com/newarchives/1016p26.shtml.

32. Mozaffarian D. Dietary and Policy Priorities for Cardiovascular Disease, Diabetes, and Obesity: A Comprehensive Review. *Circulation.* 2016;133(2):187-225. doi:10.1161/CIRCULATIONAHA.115.018585.

33. Bonjour, JP. Dietary Protein: An Essential Nutrient For Bone Health. *J Am Coll Nutr.* 2005;24(6):526S-536S. doi:10.1080/07315724.2005.10719501.

34. Wang H, Livingston KA, Fox CS, Meigs JB, Jacques PF. Yogurt consumption is associated with better diet quality and metabolic profile in American men and women. *Nutr Res.* 2013;33(1):18-26. doi:10.1016/j.nutres.2012.11.009.

35. Mozaffarian D, Hao T, Rimm EB, Willett WC, Hu FB. Changes in Diet and Lifestyle and Long-Term Weight Gain in Women and Men. *New N Engl J Med.* 2011;364:2392-2404. doi:10.1056/nejmoa1014296.

36. Schaeffer J. Dairy's Probiotic Power - A Review of the Benefits of Probiotics, the Top Sources, and What's New in the Dairy Case. *Today's Dietitian.* 2014;16(8):32. www.todaysdietitian.com/newarchives/080114p32.shtml.

    Getz L. A Healthful Dose of Bacteria - Yogurt Is the Best Probiotic Source, but Clients Do Have Other Options. *Today's Dietitian.* 2011;13(10):46. www.todaysdietitian.com/newarchives/100111p46.shtml.

Chapter 3 Boost Gut

1. Theoharides TC. On the Gut Microbiome-Brain Axis and Altruism. *Clin Ther.* 2015;37(5):937–940. doi:10.1016/j.clinthera.2015.04.003.

2. West CE, Renz H, Jenmalm MC, Kozyrskyj AL, Allen KJ, Vuillermin P, Prescott SL, et al. The Gut Microbiota and Inflammatory Noncommunicable Diseases: Associations and Potentials for Gut Microbiota Therapies. *J Allergy Clin Immunol Pract.* 2015; 135(1):3-13. doi.org/10.1016/j.jaci.2014.11.012.

    Ding R, Goh W, Wu R, Yue X, Luo X, Khine WWT, Wu J, Leeb Y. Revisit Gut Microbiota and Its Impact on Human Health and Disease. *J Food Drug Anal.* 2019;27(3)2019:623-631. doi.org/10.1016/j.jfda.2018.12.012.

3. Lam YY, Mitchell AJ, Holmes AJ, Denyer GS, Gummesson A, Caterson ID, Hunt NH, Storlien LH. Role of the gut in visceral fat inflammation and metabolic disorders. *Obesity.* 2011;19(11):2113-2120. doi:10.1038/oby.2011.68.

4. Ding S, Lund PK. Role of intestinal inflammation as an early event in obesity and insulin resistance. *Curr Opin Clin Nutr Metab Care.* 2011;14(4):328. doi:10.1097/MCO.0b013e3283478727.

    Caricilli AM, Saad MJA. The role of gut microbiota on insulin resistance. *Nutrients.* 2013;5(3):829-851. doi:10.3390/nu5030829.

5. Gianchecchi E, Fierabracci A. Recent Advances on Microbiota Involvement in the Pathogenesis of Autoimmunity. *Int J Mol Sci.* 2019;20(2):283. doi:10.3390/ijms20020283.

    Mariño E, et al. Gut microbial metabolites limit the frequency of autoimmune T cells and protect against type 1 diabetes. *Nat Immunol.* 2017;18:552-562. doi:10.1038/ni.3713.

    Brahe LK, Astrup A, Larsen LH. Is butyrate the link between diet, intestinal microbiota and obesity-related metabolic diseases? *Obes Rev.* 2013;14(12):950-59. doi:10.1111/obr.12068.

6. Everard A, Cani P. Diabetes, obesity and gut microbiota. *Best Pract Res Clin Gastroenterol.* 2013;27(1):73-83. doi:10.1016/j.bpg.2013.03.007.

    Aw W, Fukuda S. Understanding the role of the gut ecosystem in diabetes mellitus. *J Diabetes Investig.* 2018;9(1):5-12. doi:10.1111/jdi.12673.

7. Caricilli AM, Saad MJA. The role of gut microbiota on insulin resistance. *Nutrients.* 2013;5(3):829-851. doi:10.3390/nu5030829.

    Brahe LK, Astrup A, Larsen LH. Is butyrate the link between diet, intestinal microbiota and obesity-related metabolic diseases? *Obes Rev.* 2013;14(12):950-59. doi:10.1111/obr.12068.

8. Turnbaugh PJ, Hamady M, Yatsunenko T, Cantarel BL, Duncan A, Ley RE, Sogin ML, Jones WJ, Roe BA, Affourtit JP, Egholm M, Henrissat B, Heath AC, Knight R, Gordon JI. A core gut microbiome in obese and lean twins. *Nature.* 2009;457(7228):480-4. doi:10.1038/nature07540.

9. Krajmalnik-Brown R, Ilhan ZE, Kang DW, DiBaise JK. Effects of gut microbes on nutrient absorption and energy regulation. *Nutr Clin Pract.* 2012;27(2):201-214. doi:10.1177/0884533611436116.

10. Traoret CJ, Lokko P, Cruz AC, Oliveira CG, Costa NM, Bressan J, Alfenas RC, Mattes RD. Peanut digestion and energy balance. *Int J Obes (Lond).* 2008;32(2):322-8. doi:10.1038/sj.ijo.0803735.

11. Krajmalnik-Brown R, Ilhan ZE, Kang DW, DiBaise JK. Effects of gut microbes on nutrient absorption and energy regulation. *Nutr Clin Pract.* 2012;27(2):201-214. doi:10.1177/0884533611436116.

12. Krajmalnik-Brown R, Ilhan ZE, Kang DW, DiBaise JK. Effects of gut microbes on nutrient absorption and energy regulation. *Nutr Clin Pract.* 2012;27(2):201-214. doi:10.1177/0884533611436116.

13. Mozaffarian D, Hao T, Rimm EB, Willett WC, Hu FB. Changes in Diet and Lifestyle and Long-Term Weight Gain in Women and Men. *New N Engl J Med.* 2011;364:2392-2404. doi:10.1056/nejmoa1014296.

14. Paoli A, Mancin L, Bianco A, Thomas E, Mota JF, Piccini F. Ketogenic Diet and Microbiota: Friends or Enemies? *Genes (Basel).* 2019;10(7):534. doi: 10.3390/genes10070534.

    Heiman ML, Greenway FL. A Healthy Gastrointestinal Microbiome Is Dependent on Dietary Diversity. *Mol Metab.* 2016;5(5):317–320. doi:10.1016/j.molmet.2016.02.005.

    Krajmalnik-Brown R, Ilhan ZE, Kang DW, DiBaise JK. Effects of gut microbes on nutrient absorption and energy regulation. *Nutr Clin Pract.* 2012;27(2):201-214. doi:10.1177/0884533611436116.

    Hills RD Jr, et al. Gut Microbiome: Profound Implications for Diet and Disease. *Nutrients.* 2019;11(7):1613. doi:10.3390/nu11071613.

15. Lombardo NE. Nutrition's Potency to Help or Hurt Brain Health & Our Challenge. Talk presented at:

Food & Nutrition Conference & Expo 2016; Boston MA.

16 Selhub EM, Logan AC, Bested AC. Fermented foods, microbiota, and mental health: ancient practice meets nutritional psychiatry. *J Physiol Anthropol*. 2014;33(1):2. doi:10.1186/1880-6805-33-2.

Holscher H. The Gut-Brain Highway: Can Traffic Be Regulated by Diet? Talk presented at: Food & Nutrition Conference & Expo 2016; Boston MA.

Mayer EA, Knight R, Mazmanian SK, Cryan JF, Tillisch K. Gut microbes and the brain: paradigm shift in neuroscience. *J Neurosci*. 2014;34(46):15490-15496. doi:10.1523/JNEUROSCI.3299-14.2014.

Foster JA, Lyte M, Meyer E, Cryan JF. Gut microbiota and brain function: an evolving field in neuroscience. *Int J Neuropsychopharmacol*. 2016;19(5):pyv114. doi:10.1093/ijnp/pyv114.

Carabotti M, Scirocco A, Maselli MA, Severi C. The gut-brain axis: interactions between enteric microbiota, central and enteric nervous systems. *Ann Gastroenterol*. 2015;28(2):203. www.ncbi.nlm.nih.gov/pubmed/25830558.

17 Champeau R. Changing gut bacteria through diet affects brain function, UCLA study shows. newsroom.ucla.edu/releases/changing-gut-bacteria-through-245617. Published May 28, 2013. Accessed February 25, 2020.

18 Santa Cruz J. The Gut Microbiome's Link to Celiac Disease. *Today's Dietitian*. 2019;21(5):24. www.todaysdietitian.com/newarchives/0519p24.shtml.

Burkhart A. The Microbiome and Celiac Disease: A Bacterial Connection. Amy Burkhart, MD, RD & SMS. theceliacmd.com/microbiome-celiac-disease-bacterial-connection. Published October 28, 2019.

Breining, G. Recruiting Microbes to Fight Autoimmune Diseases. discoverysedge.mayo.edu/2019/01/14/recruiting-microbes-to-fight-auto-immune-diseases. Published January 2019. Accessed January 25, 2020.

19 Rothschild D, Weissbrod O, Barkan E, Kurilshikov A, Korem T, Zeevi D, Costea PI, Godneva A, Kalka IN, Bar N, et al. Environment dominates over host genetics in shaping human gut microbiota. *Nature*. 2018;555:210-215. doi: 10.1038/nature25973.

20 Bengmark S. Gut microbiota, immune development and function. *Pharmacol Res*. 2013:69(1):87-113. doi:10.1016/j.phrs.2012.09.002.

21 Santa Cruz J. The Gut Microbiome's Link to Celiac Disease. *Today's Dietitian*. 2019;21(5):24. www.todaysdietitian.com/newarchives/0519p24.shtml.

Baumler MD. Gut Bacteria. *Today's Dietitian*. 2013;15(6):46. www.todaysdietitian.com/newarchives/060113p46.shtml.

22 Melini F, et al. Health-Promoting Components in Fermented Foods: An Up-to-Date Systematic Review. *Nutrients*. 2019;11(5):1189. doi:10.3390/nu11051189.

23 Melini F, et al. Health-Promoting Components in Fermented Foods: An Up-to-Date Systematic Review. *Nutrients*. 2019;11(5):1189. doi:10.3390/nu11051189.

24 Getz L. A Healthful Dose of Bacteria - Yogurt Is the Best Probiotic Source, but Clients Do Have Other Options. *Today's Dietitian*. 2011;13(10):46. www.todaysdietitian.com/newarchives/100111p46.shtml.

25 Collins SC. Entering the World of Prebiotics - Are They a Precursor to Good Gut Health? *Today's Dietitian*. 2014;16(12):12. www.todaysdietitian.com/newarchives/120914p12.shtml.

26 Sun J, Ma H, Seeram NP, Rowley DC. Detection of Inulin, a Prebiotic Polysaccharide, in Maple Syrup. *J Agric Food Chem*. 2016;64(38):7142-7147. doi:10.1021/acs.jafc.6b03139.

27 Fruits and Vegetables Improve Gut Bacteria. Physicians Committee for Responsible Medicine. www.pcrm.org/news/health-nutrition/fruits-and-vegetables-improve-gut-bacteria. Published June 27, 2018. Accessed October 10, 2019.

28 Klimenko NS, Tyakht AV, Popenko AS, et al. Microbiome responses to an uncontrolled short-term diet intervention in the Frame of the Citizen Science Project. *Nutrients*. 2018;10(5):E576. doi: 10.3390/nu10050576.

Chapter 4 Push Produce

1 Lampe JW. Health effects of vegetables and fruit: assessing mechanisms of action in human experimental studies. *Am J Clin Nutr*. 1999;70(3 Suppl):475S-490S. doi:10.1093/ajcn/70.3.475s.

Aune D, Giovannucci E, Boffetta P, Fadnes LT, Keum N, Norat T, Greenwood DC, Riboli E, Vatten LJ, Tonstad S. Fruit and vegetable intake and the risk of cardiovascular disease, total cancer and all-cause mortality–a systematic review and dose-response meta- analysis of prospective studies. *Int J Epidemiol*. 2017;46(3):1029-1056. doi:10.1093/ije/dyw319.

2 Ebbeling CB, Swain JF, Feldman HA, Wong WW, Hachey DL, Garcia-Lago E, Ludwig DS. Effects of Dietary Composition During Weight Loss Maintenance: A Controlled Feeding Study. *JAMA*. 2012;307(24):2627–2634. doi:10.1001/jama.2012.6607.

Liu RH. Health benefits of fruit and vegetables are from additive and synergistic combinations of phytochemicals. *Am J Clin Nutr*. 2003 Sep;78(3 Suppl):517S-520S. doi: 10.1093/ajcn/78.3.517S.

Bhupathiraju SN, Tucker KL, Katherine L. Greater variety in fruit and vegetable intake is associated with lower inflammation in Puerto Rican adults. *Am J Clin Nutr*. 2010:93(1):37-46. doi:10.3945/ajcn.2010.29913.

Bhupathiraju SN, Wedick NM, Pan A, Manson JE, Rexrode KM, Willett WC, Rimm EB, Hu FB. Quantity and variety in fruit and vegetable intake and risk of coronary heart disease. *Am J Clin Nutr*. 2013;98(6):1514-523. doi:10.3945/ajcn.113.066381.

3 Liu RH. Health benefits of fruit and vegetables are from additive and synergistic combinations of phytochemicals. *Am J Clin Nutr*. 2003 Sep;78(3 Suppl):517S-520S. doi: 10.1093/ajcn/78.3.517S.

4 Hills RD Jr, et al. Gut Microbiome: Profound Implications for Diet and Disease. *Nutrients*. 2019;11(7):1613. doi:10.3390/nu11071613.

5 Heiman ML, Greenway FL. A Healthy Gastrointestinal Microbiome Is Dependent on Dietary Diversity. *Mol Metab*. 2016;5(5):317–320. doi: 10.1016/j.molmet.2016.02.005.

Hills RD Jr, et al. Gut Microbiome: Profound Implications for Diet and Disease. *Nutrients*. 2019;11(7):1613. doi:10.3390/nu11071613.

6. Ding RX, Goh WR, Wu RN, Yue XQ, Luo X, Khine WWT, Wu JR, Lee YK. Revisit Gut Microbiota and Its Impact on Human Health and Disease. *J Food Drug Anal.* 2019;27(3):623-631. doi:10.1016/j.jfda.2018.12.012.

7. Boyer J, Liu RH. Apple phytochemicals and their health benefits. *Nutr J.* 2004;3(1):5. doi:10.1186/1475-2891-3-5.

8. Liu RH. Health benefits of fruit and vegetables are from additive and synergistic combinations of phytochemicals. *Am J Clin Nutr.* 2003 Sep;78(3 Suppl):517S-520S. doi: 10.1093/ajcn/78.3.517S.

9. McDougall GJ, Stewart D. The Inhibitory Effects of Berry Polyphenols on Digestive Enzymes. *BioFactors.* 2005;23(4)189-195. doi:10.1002/biof.5520230403.

10. Martin A, Cherubini A, Andres-Lacueva C, Paniagua, Joseph J. Effects of fruits and vegetables on levels of vitamins E and C in the brain and their association with cognitive performance. *J Nutr Health Aging.* 2002;6(6):392-404. www.ncbi.nlm.nih.gov/pubmed/12459890.

11. Team led by Salk researchers. Link between vitamin A and learning abilities established by. Salk Institute for Biological Studies. www.salk.edu/news-release/link-between-vitamin-a-and-learning-abilities-established-by-team-led-by-salk-researchers/. Published December 22, 1998. Accessed October 10, 2019.

12. Whyte AR, Schafer G, Williams CM. Cognitive effects following acute wild blueberry supplementation in 7-to 10-year-old children. *Eur J Nutr.* 2016;55:2151. doi:10.1007/s00394-015-1029-4.

    Miller MG, Shukitt-Hale B. Berry fruit enhances beneficial signaling in the brain. *J Agric Food Chem.* 2012;60(23):5709-15. doi:10.1021/jf2036033.

13. Kang JH, Ascherio A, Grodstein F. Fruit and vegetable consumption and cognitive decline in aging women. *Ann Neurol.* 2005;57(5):713-20. doi:10.1002/ana.20476.

14. Morris MC, Wang Y, Barnes LL, Bennett DA, Dawson-Hughes B, Booth SL. Nutrients and bioactives in green leafy vegetables and cognitive decline: Prospective study. *Neurology.* 2018;90(3):e214-e222. doi: 10.1212/WNL.0000000000004815.

15. Hall, et al. Ultra-Processed Diets Cause Excess Calorie Intake and Weight Gain: An Inpatient Randomized Controlled Trial of Ad Libitum Food Intake. *Cell Metab.* 2019;30(1)67-77. doi:10.1016/j.cmet.2019.05.008.

16. Mozaffarian D, Hao T, Rimm EB, Willett WC, Hu FB. Changes in Diet and Lifestyle and Long-Term Weight Gain in Women and Men. *N Engl J Med.* 2011;364(25):2392-404. doi:10.1056/nejmoa1014296.

17. Desjardins Y. (2014) Fruit and Vegetables and Health: An Overview. In: Dixon G., Aldous D. (eds) Horticulture: Plants for People and Places. Springer, Dordrecht; 2014:965-1000.

18. Shivappa N, Steck SE, Hurley TG, Hussey JR, Ma Y, Ockene IS, Tabung F, Hebert JR. A population-based dietary inflammatory index predicts levels of C-reactive protein in the Seasonal Variation of Blood Cholesterol Study (SEASONS). *Public Health Nutr.* 2014;17(8):1825-833. doi: 10.1017/S1368980013002565.

    Shivappa N. *Dietary Inflammatory Index and its relationship with inflammation, metabolic biomarkers and mortality* [dissertation]. Columbia, South Carolina: University of South Carolina; 2014.

19. Spada PDS, Nunes de Souza GG, Bortolini GV, Henriques JAP, Salvador M. Antioxidant, mutagenic, and antimutagenic activity of frozen fruits. *J Med Food.* 2008;11(1):144-151. doi:10.1089/jmf.2007.598.

    Thalheimer JC. Treasures of Frozen Produce. *Today's Dietitian.* 2015;17(11):30. www.todaysdietitian.com/newarchives/1115p30.shtml.

    Bouzari A, Holstege D, Barrett DM. Vitamin retention in eight fruits and vegetables: a comparison of refrigerated and frozen storage. *J Agric Food Chem.* 2015;63(3):957-62. doi: 10.1021/jf5058793.

20. Beans and peas are unique foods. USDA ChooseMyPlate.gov. www.choosemyplate.gov/eathealthy/vegetables/vegetables-beans-and-peas. Accessed January 12, 2016.

21. Sun Y, Liu B, Snetselaar LG, Robinson JG, Wallace RB, Peterson LL, Bao W. Association of Fried Food Consumption with All Cause, Cardiovascular, and Cancer Mortality: Prospective Cohort Study. *BMJ.* 2019;364. doi: doi.org/10.1136/bmj.k5420.

22. Smith-Spangler C, Brandeau ML, Hunter GE, Bavinger JC, Pearson M, Eschbach PJ, Sundaram V, et al. Are organic foods safer or healthier than conventional alternatives?: a systematic review. *Ann Intern Med.* 2012;157(5):348-66. doi:10.7326/0003-4819-157-5-201209040-00007.

23. Thalheimer J. The Organic Foods Debate — Are They Healthier Than Conventional? *Today's Dietitian.* 2013;15(7):28. www.todaysdietitian.com/newarchives/070113p28.shtml.

24. Yeager D. The Nutrition Facts Label. *Today's Dietitian.* 2014;16(7):44. www.todaysdietitian.com/newarchives/070114p44.shtml.

25. 2018 World Hunger and Poverty Facts and Statistics. World Hunger News. Updated May 25, 2018. Accessed January 25, 2020.

26. Giampietro M. Sustainability and technological development in agriculture. *BioScience.* 1994;44(10):677-689. www.jstor.org/stable/1312511?seq=1.

27. Wendel J, Entine J. With 2000+ global studies affirming safety, GM foods among most analyzed subjects in science. Genetic Literacy Project. geneticliteracyproject.org/2013/10/08/with-2000-global-studies-confirming-safety-gm-foods-among-most-analyzed-subject-in-science/. Published October 8, 2013.

28. Hsaio J, Lion K. GMOs and Pesticides: Helpful or Harmful? Science in the News. sitn.hms.harvard.edu/flash/2015/gmos-and-pesticides/. Published August 10, 2015.

29. Carman JA, Vlieger HR, Ver Steeg LJ, Sneller VE, Robinson GW, Clinch-Jones CA, Haynes JI, Edwards JW. A long-term toxicology study on pigs fed a combined genetically modified (GM) soy and GM maize diet. *J Org Syst.* 2013;8(1):38-54. pdfs.semanticscholar.org/f359/b06ba7673c4c37f7cf-c77660469740f42c2b.pdf#page=38.

30. Snell C, Bernheim A, Bergé JB, Kuntz M, Pascal G, Paris A, Ricroch AE. Assessment of the health impact of GM plant diets in long-term and multigenera-

tional animal feeding trials: a literature review. *Food Chem Toxicol.* 2012;50(3-4):1134-48. doi:10.1016/j.fct.2011.11.048.

31. Lajolo FM, Genovese MI. Nutritional Significance of Lectins and Enzyme Inhibitors from Legumes. *J Agric Food Chem.* 2002;50(22):6592-6598. doi:10.1021/jf020191k.

32. Thalheimer JC. The Top 5 Soy Myths. *Today's Dietitian.* 2014;16(4):52. www.todaysdietitian.com/newarchives/040114p52.shtml.

    McCullough M. The Bottom Line on Soy and Breast Cancer Risk. American Cancer Society. www.cancer.org/cancer/news/expertvoices/post/2012/08/02/the-bottom-line-on-soy-and-breast-cancer-risk.aspx. Published August 2, 2012.

Chapter 5 Eat Early

1. Voigt RM, Forsyth CB, Green SJ, Engen PA, Keshavarzian A. Circadian Rhythm and the Gut Microbiome. *Int Rev Neurobiol.* 2016;131:193-205. doi:10.1016/bs.irn.2016.07.002.

2. Patterson RE, et al. Intermittent Fasting and Human Metabolic Health. *J Acad Nutr Diet.* 2015; 115(8):1203-12. doi:10.1016/j.jand.2015.02.018.

3. Cho H, Zhao X, Hatori M, Ruth TY, Barish GD, Lam MT, Chong LW, et al. Regulation of circadian behaviour and metabolism by REV-ERB-[agr] and REV-ERB-[bgr]. *Nature.* 2012;485(7396):123-127. doi:10.1038/nature11048.

4. Oike H, Oishi K, Kobori M. Nutrients, clock genes, and chrononutrition. Curr Nutr Rep. 2014;3(3):204-212. doi:10.1007/s13668-014-0082-6.

5. Leman, C. Could Intermittent Fasting Be The Answer to Reducing Breast Cancer Recurrence Risk? Dam. Mad. About Breast Cancer. dammadaboutbreastcancer.com/dont-read-youre-nighttime-eater/. Published October 2017. Updated October 2018. Accessed October 11, 2018.

6. Boden G, Ruiz J, Urbain JL, Chen X. Evidence for a Circadian Rhythm of Insulin Secretion. *Am J Physiol.* 1996;271(2 Pt 1):E246-52. doi:0.1152/ajpendo.1996.271.2.E246.

7. Leman, C. Could Intermittent Fasting Be The Answer to Reducing Breast Cancer Recurrence Risk? Dam. Mad. About Breast Cancer. dammadaboutbreastcancer.com/dont-read-youre-nighttime-eater/. Published October 2017. Updated October 2018. Accessed October 11, 2018.

8. Tuomi T, Nagorny C, Wierup N, Groop L, Mulder H. Increased Melatonin Signaling Is a Risk Factor for Type 2 Diabetes. *Cell Metab.* 2016;23(6):1067-1077. doi:10.1016/j.cmet.2016.04.009.

9. Bo S, Musso G, Beccuti G, Fadda M, Fedele D, Gambino R, Gentile L, Durazzo M, Ghigo E, Cassader M. Consuming more of daily caloric intake at dinner predisposes to obesity. A 6-year population-based prospective cohort study. *PLoS One.* 2014;9(9):e108467. doi:10.1371/journal.pone.0108467.

10. Garaulet M, Gómez-Abellán P, Alburquerque-Béjar JJ, Lee YC, Ordovás JM, Scheer FAJL. Timing of food intake predicts weight loss effectiveness. *Int J Obes (Lond).* 2013;37(4):604-11. doi:10.1038/ijo.2012.229.

11. Jakubowicz D, Barnea M, Wainstein J, Froy O. High caloric intake at breakfast vs. dinner differentially influences weight loss of overweight and obese women. *Obesity (Silver Spring).* 2013;21(12):2504-12. doi:10.1002/oby.20460.

12. McHill AW, Phillips AJK, Czeisler CA, Keating L, Yee K, Barger LK, Garaulet M, Scheer FAJL, Klerman EB. Later Circadian Timing of Food Intake Is Associated with Increased Body Fat. *Am J Clin Nutr.* 2017;106(5):1213-1219. doi:10.3945/ajcn.117.161588.

13. The Endocrine Society. Eating later in the day may be associated with obesity. ScienceDaily. www.sciencedaily.com/releases/2019/03/190323145204.htm. Published 23 March 2019.

14. Bo S, Musso G, Beccuti G, Fadda M, Fedele D, Gambino R, Gentile L, Durazzo M, Ghigo E, Cassader M. Consuming more of daily caloric intake at dinner predisposes to obesity. A 6-year population-based prospective cohort study. *PLoS One.* 2014;9(9):e108467. doi:10.1371/journal.pone.0108467.

15. Sutton EF, Early KS, Cefalu WT, Courtney. Early Time-Restricted Feeding Improves Insulin Sensitivity, Blood Pressure, and Oxidative Stress Even without Weight Loss in Men with Prediabetes. *Cell Metab.* 2018;27(6):1212–1221.e3. doi:10.1016/j.cmet.2018.04.010.

    Marinac CR, Nelson SH, Breen CI, Hartman SJ, Natarajan L, Pierce JP, Flatt SW, Sears DD, Patterson RE. Prolonged Nightly Fasting and Breast Cancer Prognosis. *JAMA Oncol.* 2016;2(8):1049-55. doi:10.1001/jamaoncol.2016.0164.

    Van Cauter E, Polonsky KS, Scheen AJ. Roles of Circadian Rhythmicity and Sleep in Human Glucose Regulation. *Endocr Rev.* 1997;18(5):716-38. doi:10.1210/edrv.18.5.0317.

16. Asher G, Sassone-Corsi P. Time for Food: the Intimate Interplay between Nutrition, Metabolism, and the Circadian Clock. *Cell.* 2015;161(1):84-92. doi:10.1016/j.cell.2015.03.015.

17. Ballon A, Neuenschwander M, Schlesinger S. Breakfast Skipping Is Associated with Increased Risk of Type 2 Diabetes among Adults: A Systematic Review and Meta-Analysis of Prospective Cohort Studies. *J Nutr.* 2019; 149(1):106–113. doi:10.1093/jn/nxy194.

18. Nas A, Mirza N, Franziska H, Kahlhöfer J, Keller J, et al. Impact of Breakfast Skipping Compared with Dinner Skipping on Regulation of Energy Balance and Metabolic Risk. *Am J Clin Nutr.* 2017;105(6):1351–1361. doi:10.3945/ajcn.116.151332.

19. Jakubowicz D, Froy O, Wainstein J, Boaz M. Meal timing and composition influence ghrelin levels, appetite scores and weight loss maintenance in overweight and obese adults. *Steroids.* 2012;77(4):323-31. doi:10.1016/j.steroids.2011.12.006.

20. Leidy HJ, Bossingham MJ, Mattes RD, Campbell WW. Increased dietary protein consumed at breakfast leads to an initial and sustained feeling of fullness during energy restriction compared to other meal times. *Br J Nutr.* 2009;101(6):798-803. doi:10.1017/s0007114508051532.

21. Leidy HJ, Ortinau LC, Douglas SM, Hoertel HA. Beneficial effects of a higher-protein breakfast on the appetitive, hormonal, and neural signals controlling energy intake regulation in overweight/obese,"breakfast-skipping," late-adolescent girls. *Am J Clin Nutr.* 2013;97(4):677-88. doi:10.3945/ajcn.112.053116.

Leidy HJ, Racki EM. The addition of a protein-rich breakfast and its effects on acute appetite control and food intake in 'breakfast-skipping' adolescents. *Int J Obes (Lond)*. 2010;34(7):1125-33. doi:10.1038/ijo.2010.3.

Leidy HJ, Bossingham MJ, Mattes RD, Campbell WW. Increased dietary protein consumed at breakfast leads to an initial and sustained feeling of fullness during energy restriction compared to other meal times. *Br J Nutr*. 2009;101(6):798-803. doi:10.1017/s0007114508051532.

Gwin JA, Leidy HJ. A Review of the Evidence Surrounding the Effects of Breakfast Consumption on Mechanisms of Weight Management. *Adv Nutr*. 2018;9(6):717-725. doi:10.1093/advances/nmy047.

22. Vander Wal JS, Gupta A, Khosla P, Dhurandhar NV. Egg breakfast enhances weight loss. *Int J Obes*. 2008;32(10):1545-1551. doi:10.1038/ijo.2008.130.

23. Vollmers C, Gill S, DiTacchio L, Pulivarthy SR, Le HP, Panda S. Time of feeding and the intrinsic circadian clock drive rhythms in hepatic gene expression. *Proc Natl Acad Sci USA*. 2009;106(50):21453-8. doi:10.1073/pnas.0909591106.

24. Jakubowicz D, Barnea M, Wainstein J, Froy O. High caloric intake at breakfast vs. dinner differentially influences weight loss of overweight and obese women. *Obesity (Silver Spring)*. 2013;21(12):2504-12. doi:10.1002/oby.20460.

25. Yokoyama Y, Onishi K, Hosoda T, Amano H, Otani S, Kurozawa Y, Tamakoshi A. Skipping Breakfast and Risk of Mortality from Cancer, Circulatory Diseases and All Causes: Findings from the Japan Collaborative Cohort Study. *Yonago Acta Med*. 2016;59(1):55-60. www.ncbi.nlm.nih.gov/pubmed/27046951.

26. Sutton EF, Early KS, Cefalu WT, Courtney. Early Time-Restricted Feeding Improves Insulin Sensitivity, Blood Pressure, and Oxidative Stress Even without Weight Loss in Men with Prediabetes. *Cell Metab*. 2018;27(6):1212–1221.e3. doi:10.1016/j.cmet.2018.04.010.

27. Jakubowicz D, Wainstein J, Ahrén B, Bar-Dayan Y, Landau Z, Rabinovitz HR, Froy O. High-Energy Breakfast with Low-Energy Dinner Decreases Overall Daily Hyperglycaemia in Type 2 Diabetic Patients: a Randomised Clinical Trial. *Diabetologia*. 2015;58(5):912-9. doi:10.1007/s00125-015-3524-9.

28. Sutton EF, Early KS, Cefalu WT, Courtney. Early Time-Restricted Feeding Improves Insulin Sensitivity, Blood Pressure, and Oxidative Stress Even without Weight Loss in Men with Prediabetes. *Cell Metab*. 2018;27(6):1212–1221.e3. doi:10.1016/j.cmet.2018.04.010.

29. Hispanic Community Health Study/Study of Latinos (HCHS/SOL). www.nhlbi.nih.gov/science/hispanic-community-health-studystudy-latinos-hchssol. Accessed January 25, 2020.

30. Hall et al. Ultra-Processed Diets Cause Excess Calorie Intake and Weight Gain: An Inpatient Randomized Controlled Trial of Ad Libitum Food Intake. *Cell Metab*. 2019;30(2):67–77. doi:10.1016/j.cmet.2019.05.008.

31. Veldhorst MAB, Westerterp-Plantenga MS, Westerterp KR. Gluconeogenesis and Energy Expenditure after a High-Protein, Carbohydrate-Free Diet. *Am J Clin Nutr*. 2009;90(3):519-26. doi:10.3945/ajcn.2009.27834.

32. Wallace TC, et al. Fruits, Vegetables, and Health: A Comprehensive Narrative, Umbrella Review of the Science and Recommendations for Enhanced Public Policy to Improve Intake. *Crit Rev Food Sci Nutr*. (2019). doi:10.1080/10408398.2019.1632258.

Zong G, Gao A, Hu FB, Sun Q. Whole Grain Intake and Mortality From All Causes, Cardiovascular Disease, and Cancer: A Meta-Analysis of Prospective Cohort Studies. *Circulation*. 2016;133(24):2370-80. doi:10.1161/CIRCULATIONAHA.

33. Slavin JL, Lloyd B. Health benefits of fruits and vegetables. *Adv Nutr*. 2012;3(4):506-16. doi:10.3945/an.112.002154.

34. Vanegas SM, Meydani M, Barnett JB, Goldin B, Kane A, Rasmussen H, Brown C, et al. Substituting whole grains for refined grains in a 6-wk randomized trial has a modest effect on gut microbiota and immune and inflammatory markers of healthy adults. *Am J Clin Nutr*. 2017;105(3):635-650. doi:10.3945/ajcn.116.146928.

Heiman ML, Greenway FL. A Healthy Gastrointestinal Microbiome Is Dependent on Dietary Diversity. *Mol Metab*. 2016;5(5):317–320. doi:10.1016/j.molmet.2016.02.005.

35. Masood W. Ketogenic Diet. *StatPearls [Internet]*. (2019). www.ncbi.nlm.nih.gov/pubmed/29763005.

36. Low-Carb Diet Tied to Common Heart Rhythm Disorder. ScienceDaily. www.sciencedaily.com/releases/2019/03/190306081652.htm. Published March 6, 2019.

37. Bergqvist AG, Schall JI, Stallings VA, Zemel BS. Progressive bone mineral content loss in children with intractable epilepsy treated with the ketogenic diet. *Am J Clin Nutr*. 2008;88(6):1678-84. doi:10.3945/ajcn.2008.26099.

38. Hall et al. Ultra-Processed Diets Cause Excess Calorie Intake and Weight Gain: An Inpatient Randomized Controlled Trial of Ad Libitum Food Intake. *Cell Metab*. 2019;30(2):67–77. doi:10.1016/j.cmet.2019.05.008.

39. Masood W. Ketogenic Diet. *StatPearls [Internet]*. (2019). www.ncbi.nlm.nih.gov/pubmed/29763005.

40. White AM, Johnston CS, Swan PD, Tjonn SL, Sears B. Blood Ketones Are Directly Related to Fatigue and Perceived Effort during Exercise in Overweight Adults Adhering to Low-Carbohydrate Diets for Weight Loss: a Pilot Study. *J Am Diet Assoc*. 2007;107(10):1792-6. doi:10.1016/j.jada.2007.07.009.

41. U.S. News Reveals Best Diets Rankings for 2018. U.S. News & World Report. www.usnews.com/info/blogs/press-room/articles/2018-01-03/us-news-reveals-best-diets-rankings-for-2018. Published January 3, 2018. Accessed October 20, 2019.

42. Paoli A, Mancin L, Bianco A, Thomas E, Mota JF, Piccini F. Ketogenic Diet and Microbiota: Friends or Enemies? *Genes (Basel)*. 2019;10(7):534. doi: 10.3390/genes10070534.

43. Paoli A, Mancin L, Bianco A, Thomas E, Mota JF, Piccini F. Ketogenic Diet and Microbiota: Friends or Enemies? *Genes (Basel)*. 2019;10(7):534. doi: 10.3390/genes10070534.

Krajmalnik-Brown R, Ilhan ZE, Kang DW, DiBaise JK. Effects of gut microbes on nutrient absorption and energy regulation. *Nutr Clin Pract*. 2012;27(2):201-214. doi:10.1177/0884533611436116.

Hills RD Jr, et al. Gut Microbiome: Profound Implications for Diet and Disease. *Nutrients.* 2019;11(7):1613. doi:10.3390/nu11071613.

44 University of Alabama at Birmingham. Time-Restricted Feeding Study Shows Promise in Helping People Shed Body Fat. Science Daily. www.sciencedaily.com/releases/2017/01/170106113820.htm. Published January 6, 2017. Accessed October 20, 2019.

45 Gill S, Panda S. A Smartphone App Reveals Erratic Diurnal Eating Patterns in Humans That Can Be Modulated for Health Benefits. *Cell Metab.* 2015;22(5):789-98. doi:10.1016/j.cmet.2015.09.005.

46 Anton S. Debate: Intermittent Fasting in Weight Management. Presented at FNCE; October 21, 2018; Chicago.

Chapter 6 Move More

American Heart Association Recommendations for Physical Activity in Adults. American Heart Association. atgprod.heart.org/HEARTORG/HealthyLiving/PhysicalActivity/StartWalking/American-Heart-Association-Recommendations-for-Physical-Activity-in-Adults_UCM_307976_Article.jsp. Published February 2014. Accessed April 27, 2017.

Garber CE, Blissmer B, Deschenes MR, Franklin BA, Lamonte MJ, Lee IM, Nieman DC, Swain DP. Quantity and Quality of Exercise for Developing and Maintaining Cardiorespiratory, Musculoskeletal, and Neuromotor Fitness in Apparently Healthy Adults. *Med Sci Sports Exerc.* 2011;43(7):1334-359. doi:10.1249/mss.0b013e318213fefb.

American College of Sports Medicine. ACSM Quantity and Quality of Exercise for Developing and Maintaining Cardiorespiratory, Musculoskeletal, and Neuromotor Fitness in Apparently Healthy Adults: Guidance for Prescribing Exercise. *Med Sci Sports Exerc.* 2011;43(7):1334-59. doi:10.1249/MSS.0b013e318213fefb.

How much physical activity do adults need? Centers for Disease Control and Prevention. www.cdc.gov/physicalactivity/basics/adults/index.htm. Published June 04, 2015. Updated January 9, 2020.

Delavier F. *Women's Strength Training Anatomy.* Champaign, IL: Human Kinetics; 2003.

Giovanni I, Rigutti E. *Atlas of Anatomy.* Harvard International: Taj Books Ltd; 2005.

Clark MA, Lucett S, Corn RJ. *NASM Essentials of Personal Fitness Training.* 3rd ed. Philadelphia: Lippincott Williams & Wilkins; 2008.

Staugaard-Jones JA. *The Concise Book of Yoga Anatomy: An Illustrated Guide to the Science of Motion.* Berkeley: North Atlantic Books; 2015.

Folkins CH, Sime WE. Physical fitness training and mental health. *Am Psychol.* 1981;36(4):373–389. doi:10.1037/0003-066X.36.4.373.

Pescatello L, Arena R, Riebe D, Thompson PD, American College of Sports Medicine. *ACSM's Guidelines for Exercise Testing and Prescription.* 9th ed. Philadelphia, PA: Lippincott Williams & Wilkins; 2013.

Layman DK. Protein nutrition, meal timing, and muscle health. In: Berdanier CD, Dwyer JT, Heber D, ed. *Handbook of Nutrition and Food.* 3rd ed. Boca Raton, FL: CRC Press; 2013:861-868.

1 Wang S, Luo X, Barnes D, Sano M, Yaffe K. Physical activity and risk of cognitive impairment among oldest-old women. *Am J Geriatr Psychiatry.* 2014;22(11):1149-57. doi:10.1016/j.jagp.2013.03.002.

2 Nagamatsu LS, Chan A, Davis JC, Beattie BL, Graf P, Voss MW, Sharma D, Liu-Ambrose T. Physical activity improves verbal and spatial memory in older adults with probable mild cognitive impairment: a 6-month randomized controlled trial. *J Aging Res.* 2013;2013:861893. doi:10.1155/2013/861893.

3 Anderson E, Shivakumar G. Effects of exercise and physical activity on anxiety. *Front Psychiatry.* 2013;4:27. doi:10.3389/fpsyt.2013.00027.

4 Herting MM, Chu X. Exercise, cognition, and the adolescent brain. *Birth Defects Res.* 2017;109(20):1672-1679. doi:10.1002/bdr2.1178.

5 Colberg SR, et al. Postprandial Walking is Better for Lowering the Glycemic Effect of Dinner than Pre-Dinner Exercise in Type 2 Diabetic Individuals. *J Am Med Dir Assoc.* 2009;10(6):394-397. doi:10.1016/j.jamda.2009.03.015.

6 Gomes da Silva S, Arida RM. Physical activity and brain development. *Expert Rev Neurother.* 2015;15(9):1041-51. doi:10.1586/14737175.2015.1077115.

7 American College of Sports Medicine, Sawka MN, Burke LM, Eichner ER, Maughan RJ, Montain SJ, Stachenfeld NS. American College of Sports Medicine position stand. Exercise and fluid replacement. *Med Sci Sports Exerc.* 2007;39(2):377-90. doi:10.1249/mss.0b013e31802ca597.

8 American College of Sports Medicine, Sawka MN, Burke LM, Eichner ER, Maughan RJ, Montain SJ, Stachenfeld NS. American College of Sports Medicine position stand. Exercise and fluid replacement. *Med Sci Sports Exerc.* 2007;39(2):377-90. doi:10.1249/mss.0b013e31802ca597.

9 Aragon AA, Schoenfeld BJ. Nutrient timing revisited: is there a post-exercise anabolic window?. *J Int Soc Sports Nutr.* 2013;10:5. doi:10.1186/1550-2783-10-5.

Burd NA, West DWD, Moore DR, Atherton PJ, Staples AW, Prior T., et al. Enhanced amino acid sensitivity of myofibrillar protein synthesis persists for up to 24 h after resistance exercise in young men. *J Nutr.* 2011;141(4):568-573. doi:10.3945/jn.110.135038.

Mamerow MM, Mettler JA, English KL, Casperson SL, Arentson-Lantz E, Sheffield-Moore M, Layman DK, Paddon-Jones D. Dietary protein distribution positively influences 24-h muscle protein synthesis in healthy adults. *J Nutr.* 2014;144(6): 876–880. doi:10.3945/jn.113.185280.

10 Thomas DT, Erdman KA, Burke LM. Position of the academy of nutrition and dietetics, dietitians of canada, and the american college of sports medicine: Nutrition and athletic performance. *J Acad Nutr Diet.* 2016;116(3):501-528. doi:10.1016/j.jand.2015.12.006.

11 Rasmussen B. Muscle and/to Brains: Understanding Protein Synthesis and Efficiency. Presented at FNCE; October 4, 2015; Nashville.

12 Clark N. Nancy Clark's Sports Nutrition Guidebook. 6th ed. Newton, MA: Sports Nutrition Services; 2020.

Chapter 7 Water Well

1 Thornton SN. Increased Hydration Can Be Associated with Weight Loss. *Front Nutr.* 2016;3:18. doi:10.3389/fnut.2016.00018.

2 Stookey JD, Constant F, Gardner CD, Popkin BM. Replacing sweetened caloric beverages with drinking

water is associated with lower energy intake. *Obesity*. 2007:15(12);3013-3022. doi:10.1038/oby.2007.359.

Schulze MB, Manson JE, Ludwig DS, Colditz GA, Stampfer MJ, Willett WC, Hu FB. Sugar-sweetened beverages, weight gain, and incidence of type 2 diabetes in young and middle-aged women. *JAMA*. 2004;292(8):927-934. doi:10.1001/jama.292.8.927.

3 Tate DF, Turner-McGrievy G, Lyons E, Stevens J, Erickson K, Polzien K, Diamond M, Wang X, Popkin B. Replacing caloric beverages with water or diet beverages for weight loss in adults: main results of the Choose Healthy Options Consciously Everyday (CHOICE) randomized clinical trial. *Am J Clin Nutr*. 2012;95(3):555-63. doi:10.3945/ajcn.111.026278.

4 Mattes RD. Hunger and thirst: issues in measurement and prediction of eating and drinking. *Physiol Behav*. 2010;100(1):22-32. doi:10.1016/j.physbeh.2009.12.026.

5 Boschmann M, Steiniger J, Hille U, Tank J, Adams F, Sharma AM, Klaus S, Luft FC, Jordan J. Water-induced thermogenesis. *J Clin Endocrinol Metab*. 2003;88(12):6015-9. doi:10.1210/jc.2003-030780.

6 Stookey JD, Constant F, Popkin BM, Gardner CD. Drinking water is associated with weight loss in overweight dieting women independent of diet and activity. *Obesity*. 2008;16(11):2481-8. doi:10.1038/oby.2008.409.

7 Parretti HM, Aveyard P, Blannin A, Clifford SJ, Coleman SJ, Roalfe A, Daley AJ. Efficacy of water preloading before main meals as a strategy for weight loss in primary care patients with obesity: RCT. *Obesity*. 2015;23(9)1785-1791. doi:10.1002/oby.21167.

8 Ganio MS, Armstrong LE, Casa DJ, McDermott BP, Lee EC, Yamamoto LM, Marzano S, et al. Mild dehydration impairs cognitive performance and mood of men. *Br J Nutr*. 2011;106(10):1535-1543. doi:10.1017/S0007114511002005.

9 Ganio MS, Armstrong LE, Casa DJ, McDermott BP, Lee EC, Yamamoto LM, Marzano S, et al. Mild dehydration impairs cognitive performance and mood of men. *Br J Nutr*. 2011;106(10):1535-1543. doi:10.1017/S0007114511002005.

Armstrong LE, Ganio MS, Casa DJ, Lee EC, McDermott BP, Klau JF, Jimenez L, Le Bellego L, Chevillotte E, Lieberman HR. Mild dehydration affects mood in healthy young women. *J Nutr*. 2012;142(2):382-388. doi:10.3945/jn.111.142000.

10 Kalman DS, Lepeley A. A review of hydration. *J Strength Cond Res*. 2010;32(2):56-63. doi:10.1519/SSC.0b013e3181c21172.

11 Madjd A, Taylor MA, Delavari A, Malekzadeh R, Macdonald IA, Farshchi HR. Beneficial Effects of Replacing Diet Beverages with Water on Type 2 Diabetic Obese Women Following a Hypo-energetic Diet: A Randomized, 24-week Clinical Trial. *Diabetes Obes Metab*. 2017;19(1):125-132. doi:10.1111/dom.12793.

12 Ruiz-Ojeda FJ, et al. Effects of Sweeteners on the Gut Microbiota: A Review of Experimental Studies and Clinical Trials. *Adv Nutr*. 2019;10(Suppl 1):S31–S48. doi:10.1093/advances/nmy037.

Chi L, Bian X, Gao B, Tu P, Lai Y, Ru H, Lu K. Effects of the Artificial Sweetener Neotame on the Gut Microbiome and Fecal Metabolites in Mice. *Molecules*. 2018;23(2);367. doi:10.3390/molecules23020367.

13 Higgins JP, Babu K, Deuster PA, Shearer J. Energy Drinks: A Contemporary Issues Paper. *Curr Sports Med Rep*. 2018;17(2);65-72. doi:10.1249/JSR.0000000000000454.

14 Sorkin, Barbara C, and Paul M Coates. Caffeine-containing energy drinks: beginning to address the gaps in what we know. *Adv Nutr*. 2014;5(5):541-543. doi:10.3945/an.114.006411.

Howland J, Rohsenow DJ. Risks of Energy Drinks Mixed With Alcohol. *JAMA*. 2013;309(3):245-246. doi:10.1001/jama.2012.187978.

15 Energy Drinks. The National Center for Complementary and Integrative Health (NCCIH). nccih.nih.gov/health/energy-drinks. Published July 26, 2018.

16 Armstrong LE, Ganio MS, Casa DJ, Lee EC, McDermott BP, Klau JF, Jimenez L, Le Bellego L, Chevillotte E, Lieberman HR. Mild dehydration affects mood in healthy young women. *J Nutr*. 2012;142(2):382-388. doi:10.3945/jn.111.142000.

17 Chakravorty S, Bhattacharya S, Chatzinotas A, Chakraborty W, Bhattacharya D, Gachhui R. Kombucha Tea Fermentation: Microbial and Biochemical Dynamics. *Int J Food Microbiol*. 2016;220(2):63–72. doi:10.1016/j.ijfoodmicro.2015.12.015.

18 SungHee Kole A, Jones HD, Christensen R, Gladstein J. A Case of Kombucha Tea Toxicity. *J Intensive Care Med*. 2009;24(3):205-7. doi:10.1177/0885066609332963.

19 Mayser P, Fromme S, Leitzmann G, Gründer K. The Yeast Spectrum of the 'Tea Fungus Kombucha.' *Mycoses*. 1995;38(7-8):289-95. doi:10.1111/j.1439-0507.1995.tb00410.x.

20 Bains C, The Canadian Press. B.C. institute tests 700 samples of kombucha for alcohol levels. The Star. www.thestar.com/vancouver/2019/11/07/bc-institute-tests-700-samples-of-kombucha-for-alcohol-levels.html. Published November 7, 2019.Accessed February 24, 2020.

21 Caballero M. Kombucha Study Raises Sugar Content Questions. Bevnet. www.bevnet.com/news/2016/kombucha-study-raises-sugar-content-questions. Published November 15, 2016.

NaturPro. Sugars in Kombucha Tea. NaturPro Scientific. http://naturproscientific.com/portfolio/sugars-in-kombucha-tea/. Accessed November 12, 2019.

22 Bourrie BCT, Willing BP, Cotter PD. The Microbiota and Health Promoting Characteristics of the Fermented Beverage Kefir. *Front Microbiol*. 2016;7:647. doi:10.3389/fmicb.2016.00647.

23 Volpi G, Ginepro M, Tafur-Marinos J, Zelano V. Pollution Abatement of Heavy Metals in Different Conditions by Water Kefir Grains as a Protective Tool against Toxicity. *Journal of Chemistry*. 2019;:1-10. doi:10.1155/2019/8763902.

24 Jiang L, Gulanski BI, De Feyter HM, Weinzimer SA, Pittman B, Guidone E, Koretski J, et al. Increased Brain Uptake and Oxidation of Acetate in Heavy Drinkers. *J Clin Invest*. 2013;123(4):1605-1614. doi:10.1172/JCI65153.

25 Cains S, Blomeley C, Kollo M, et al. Agrp neuron activity is required for alcohol-induced overeating. *Nat Commun*. 2017;8:14014. doi:10.1038/ncomms14014.

26. U.S. Department of Health and Human Services and U.S. Department of Agriculture. Appendix 9. Alcohol. In: *2015-2020 Dietary Guidelines for Americans.* 8th ed. 2015. health.gov/dietaryguidelines/2015/guidelines/appendix-9/. Accessed October 20, 2019.

27. Cains S, Blomeley C, Kollo M, et al. Agrp neuron activity is required for alcohol-induced overeating. *Nat Commun.* 2017;8:14014. doi:10.1038/ncomms14014.

28. American College of Sports Medicine, Sawka MN, Burke LM, Eichner ER, Maughan RJ, Montain SJ, Stachenfeld NS. American College of Sports Medicine position stand. Exercise and fluid replacement. *Med Sci Sports Exerc.* 2007;39(2):377-90. doi:10.1249/mss.0b013e31802ca597.

Kalman DS, Lepeley A. A review of hydration. *J Strength Cond Res.* 2010;32(2):56-63. doi:10.1519/SSC.0b013e3181c21172.

Chapter 8 Shift Surroundings

1. Wansink B, Kim J. Bad Popcorn in Big Buckets: Portion Size Can Influence Intake as Much as Taste. *J Nutr Educ Behav.* 2005;37(5):242-45. doi:10.1016/s1499-4046(06)60278-9.

2. Watkins, J.A. Mindless Eating: Why We Eat More Than We Think, Brian Wansink, Ph.D.. *J Behav Health Serv Res.* 2008;35:235–236. doi:10.1007/s11414-007-9102-2.

3. Watkins, J.A. Mindless Eating: Why We Eat More Than We Think, Brian Wansink, Ph.D.. *J Behav Health Serv Res.* 2008;35:235–236. doi:10.1007/s11414-007-9102-2.

4. Wansink B, Painter JE, Lee YK. The office candy dish: proximity's influence on estimated and actual consumption. *Int J Obes (Lond).* 2006;30(5):871-5. doi:10.1038/sj.ijo.0803217.

5. Wansink B, van Ittersum K. Portion size me: Plate-size induced consumption norms and win-win solutions for reducing food intake and waste. *J Exp Psychol Appl.* 2013;19(4):320-32. doi:10.1037/a0035053.

6. Wansink B, van Ittersum K, Painter JE. Ice Cream Illusions: Bowls, Spoons, and Self-Serve Portion Sizes. *Am J Prev Med.* 2006;31(3):240-3. doi:10.1016/j.amepre.2006.04.003.

7. Mishra A, Mishra H, Masters T. The Influence of Bite Size on Quantity of Food Consumed: A Field Study. *Journal of Consumer Research.* 2012;38(5):791–795. doi:10.1086/660838.

8. Wansink B, Van Ittersum K. Shape of glass and amount of alcohol poured: comparative study of effect of practice and concentration. *BMJ.* 2005;331(7531):1512–1514. doi:10.1136/bmj.331.7531.1512.

9. Hollands GJ, Shemilt I, Marteau TM, Jebb SA, Lewis HB, Wei Y, Higgins JP, Ogilvie D. Portion, package or tableware size for changing selection and consumption of food, alcohol and tobacco. *Cochrane Database Syst Rev.* 2015;2015(9):CD011045. doi:10.1002/14651858.CD011045.pub2.

Chapter 9 Switch Fat

1. Granholm AC, Bimonte-Nelson H, Moore AB, Nelson ME, Freeman LR, Sambamurti K. Effects of a saturated fat and high cholesterol diet on memory and hippocampal morphology in the middle-aged rat. *J Alzheimers Dis.* 2008;14(2):133-145. doi:0.3233/JAD-2008-14202.

Gardener S, Rainey-Smith S, Barnes M, et al. Dietary patterns and cognitive decline in an Australian study of ageing. *Mol Psychiatry.* 2015;20:860-866. doi:10.1038/mp.2014.79.

2. Gardener S, Rainey-Smith S, Barnes M, et al. Dietary patterns and cognitive decline in an Australian study of ageing. *Mol Psychiatry.* 2015;20:860-866. doi:10.1038/mp.2014.79.

3. Titos E, Clària J. Omega-3-derived mediators counteract obesity-induced adipose tissue inflammation. *Prostaglandins Other Lipid Mediat.* 2013;107:77-84. doi10.1016/j.prostaglandins.2013.05.003.

Calder PC. Omega-3 fatty acids and inflammatory processes. *Nutrients.* 2010;2(3):355-74. doi:10.3390/nu2030355.

4. Weisenberger J. The Omega Fats. *Today's Dietitian.* 2014;16(4):20. www.todaysdietitian.com/newarchives/040114p20.shtml.

5. Bradbury J. Docosahexaenoic acid (DHA): an ancient nutrient for the modern human brain. *Nutrients.* 2011;3(5):529-54. doi:10.3390/nu3050529.

6. Judge MP, Harel O, Lammi-Keefe CJ. Maternal consumption of a docosahexaenoic acid-containing functional food during pregnancy: benefit for infant performance on problem-solving but not on recognition memory tasks at age 9 mo. *Am J Clin Nutr.* 2007;85(6):1572-1577. doi:10.1093/ajcn/85.6.1572.

7. FAO/WHO. Report of the Joint FAO/WHO Expert Consultation on the Risks and Benefits of Fish Consumption. Rome: FAO/WHO; 2011. Published January 2011.

8. Morris MC, et al. MIND diet slows cognitive decline with aging. *Alzheimers Dement.* 2015;11(9):1015–1022. doi: 10.1016/j.jalz.2015.04.011.

9. Rush University Medical Center. Stave off cognitive decline with seafood: Study finds that eating seafood once a week may slow memory loss. ScienceDaily. www.sciencedaily.com/releases/2016/05/160510124831.htm. Published 10 May 2016.

10. The Alzheimer's Information Site. Eating Seafood Lowers Risk of Alzheimer's Brain Changes. Fisher Center for Alzheimer's Research Foundation. www.alzinfo.org/articles/prevention/eating-seafood-lowers-risk-of-alzheimers-brain-changes/. Published 6 Apr. 2016.

11. Harrar S. Omega-3 Fatty Acids and Mood Disorders. *Today's Dietitian.* 2012;1(1):22. www.todaysdietitian.com/newarchives/011012p22.shtml.

Sarris J, Mischoulon D, Schweitzer I. Omega-3 for bipolar disorder: meta-analyses of use in mania and bipolar depression. *J Clin Psychiatry.* 2012;73(1):81-6. doi:10.4088/JCP.10r06710.

12. Hibbeln JR. Seafood consumption, the DHA content of mothers' milk and prevalence rates of postpartum depression: a cross-national, ecological analysis. *J Affect Disord.* 2002:69(1):15-29. doi:10.1016/S0165-0327(01)00374-3.

13. Escamilla-Nuñez MC, Barraza-Villarreal A, Hernández-Cadena L, Navarro-Olivos E, Sly PD, Romieu I. Omega-3 fatty acid supplementation

during pregnancy and respiratory symptoms in children. *Chest*. 2014;146(2):373-382. doi:10.1378/chest.13-1432.

14 Alehagen U, Johansson P, Björnstedt M, et al. Relatively high mortality risk in elderly Swedish subjects with low selenium status. *Eur J Clin Nutr*. 2016;70:91–96. doi:10.1038/ejcn.2015.92.

Ralston N. Selenium: The Secret That Will Change Public Perception of Seafood. Talk presented at: Academy of Nutrition and Dietetics 2016 Food and Nutrition Conference & Expo; October 16, 2016; Boston, MA.

Ralston NV, Raymond LJ. Dietary selenium's protective effects against methylmercury toxicity. *Toxicology*. 2010;278(1):112-23. doi:10.1016/j.tox.2010.06.004.

15 Hoffmann PR, Berry MJ. The influence of selenium on immune responses. *Mol Nutr Food Res*. 2008;52(11):1273–1280. doi:10.1002/mnfr.200700330.

16 U.S. Food and Drug Administration. Advice about Eating Fish. www.fda.gov/food/consumers/advice-about-eating-fish. Published July 2, 2019. Accessed January 25, 2020.

17 Simopoulos AP, Salem N Jr. n-3 fatty acids in eggs from range-fed Greek chickens. *N Engl J Med*. 1989;321(20):1412. doi:10.1056/NEJM198911163212013.

18 Schwingshackl L, Strasser B, Hoffmann G. Effects of monounsaturated fatty acids on cardiovascular risk factors: a systematic review and meta-analysis. *Ann Nutr Metab*. 2011;59(2-4):176-86. doi:10.1159/000334071.

Gillingham LG, Harris-Janz S, Jones PJ. Dietary monounsaturated fatty acids are protective against metabolic syndrome and cardiovascular disease risk factors. *Lipids*. 2011;46(3):209-28. doi:10.1007/s11745-010-3524-y.

van Dijk SJ, Feskens EJM, Bos MB, de Groot LCPGM, de Vries JHM, Müller M, Afman L. Consumption of a high monounsaturated fat diet reduces oxidative phosphorylation gene expression in peripheral blood mononuclear cells of abdominally overweight men and women. *J Nutr*. 2012;142(7):1219–1225. doi:10.3945/jn.111.155283.

19 Liu X, Kris-Etherton PM, West SG, Lamarche B, Jenkins DJ, Fleming JA, McCrea CE, Pu S, Couture P, Connelly PW, Jones PJ. Effects of canola and high-oleic-acid canola oils on abdominal fat mass in individuals with central obesity. *Obesity (Silver Spring)*. 2016;24(11):2261-2268. doi:10.1002/oby.21584.

20 Lin L, Allemekinders H, Dansby A, Campbell L, Durance-Tod S, Berger A, Jones PJH. Evidence of health benefits of canola oil. *Nutr Rev*. 2013;71(6):370–385. doi:10.1111/nure.12033.

Dennett C. Healthful Oils: The Canola Controversy. *Today's Dietitian*. 2018;20(10):12. www.todaysdietitian.com/newarchives/1018p12.shtml.

21 Liu X, Kris-Etherton PM, West SG, Lamarche B, Jenkins DJ, Fleming JA, McCrea CE, Pu S, Couture P, Connelly PW, Jones PJ. Effects of canola and high-oleic-acid canola oils on abdominal fat mass in individuals with central obesity. *Obesity (Silver Spring)*. 2016;24(11):2261-2268. doi:10.1002/oby.21584.

Jones PJ, MacKay DS, Senanayake VK, Pu S, Jenkins DJ, Connelly PW, Lamarche B, et al. High-oleic canola oil consumption enriches LDL particle cholesteryl oleate content and reduces LDL proteoglycan binding in humans. *Atherosclerosis*. 2015;238(2):231-8. doi:10.1016/j.atherosclerosis.2014.12.010.

22 Gorzynik-Debicka M, et al. Potential Health Benefits of Olive Oil and Plant Polyphenols. *Int J Mol Sci*. 2018;19(3):686. doi:10.3390/ijms19030686.

Rigacci S, Stefani M. Nutraceutical Properties of Olive Oil Polyphenols. An Itinerary from Cultured Cells through Animal Models to Humans. *Int J Mol Sci*. 2016;17(6). doi:10.3390/ijms17060843.

23 Rinaldi de Alvarenga JF, Quifer-Rada P, Francetto Juliano F, Hurtado-Barroso S, Illan M, Torrado-Prat X, Lamuela-Raventós RM. Using Extra Virgin Olive Oil to Cook Vegetables Enhances Polyphenol and Carotenoid Extractability: A Study Applying the sofrito Technique. *Molecules*. 2019;24(8). doi:10.3390/molecules24081555.

24 Harvard Health Publishing. The Truth about Fats: the Good, the Bad, and the in-Between. Harvard Health. www.health.harvard.edu/staying-healthy/the-truth-about-fats-bad-and-good. Published February 2015. Accessed October 8, 2019.

25 Moran, B. Is Butter Really Back? Harvard Public Health. www.hsph.harvard.edu/magazine/magazine_article/is-butter-really-back. Published 2014. Accessed January 25, 2020.

26 We Repeat: Butter is Not Back. The Nutrition Source. www.hsph.harvard.edu/nutritionsource/2016/06/30/we-repeat-butter-is-not-back. Published June 30, 2016. Accessed January 25, 2020.

McCulloch M. Saturated Fat: Not So Bad or Just Bad Science? *Today's Dietitian*. 2014;16(11):32. www.todaysdietitian.com/newarchives/111114p32.shtml.

27 McCulloch M. Saturated Fat: Not So Bad or Just Bad Science? *Today's Dietitian*. 2014;16(11):32. www.todaysdietitian.com/newarchives/111114p32.shtml.

28 Thalheimer, J. Coconut Oil. *Today's Dietitian*. 2016;18(10):32. www.todaysdietitian.com/newarchives/1016p32.shtml.

29 Neelakantan N, Seah J, and van Dam, RM. The Effect of Coconut Oil Consumption on Cardiovascular Risk Factors: A Systematic Review and Meta-Analysis of Clinical Trials. *Circulation*. 2020;141:1-12. doi:10.1161/CIRCULATIONAHA.119.043052.

30 "Free Range" and "Pasture Raised" officially defined by HFAC for Certified Humane® label. Certified Humane. certifiedhumane.org/free-range-and-pasture-raised-officially-defined-by-hfac-for-certified-humane-label/. Published October 9, 2014. Accessed December 17, 2019.

A Consumer's Guide to Food Labels and Animal Welfare. Animal Welfare Institute. awionline.org/content/consumers-guide-food-labels-and-animal-welfare. Accessed January 25, 2020.

31 Gorman, RM. Are Cage-Free Eggs Really Better?. EatingWell. www.eatingwell.com/article/289956/are-cage-free-eggs-really-better. Published January 2017. Accessed January 25, 2020.

Chapter 10 Change Grains

1 Shivappa N. *Dietary Inflammatory Index and its relationship with inflammation, metabolic biomarkers*

2. Ozawa M, Shipley M, Kivimaki M, Singh-Manoux A, Brunner EJ. Dietary pattern, inflammation and cognitive decline: The Whitehall II prospective cohort study. *Clin Nutr.* 2017;36(2):506-512. doi:10.1016/j.clnu.2016.01.013.

3. Olga C. A Survey on dietary intake and eating habits among the students from the Republic of Moldova. *Global Journal of Multidisciplinary Studies.* 2016;5(11).

4. Slavin JL, Martini MC, Jacobs DR, Marquart L. Plausible mechanisms for the protectiveness of whole grains. *Am J Clin Nutr.* 1999:70(3):459s–463s. doi:10.1093/ajcn/70.3.459s.

   Slavin J. Why whole grains are protective: biological mechanisms. *Proc Nutr Soc.* 2003;62(1):129-134. doi:10.1079/PNS2002221.

5. Brown L, Rosner B, Willett WW, Sacks FM. Cholesterol-lowering effects of dietary fiber: a meta-analysis. *Am J Clin Nutr.* 1999;69(1):30-42. doi:10.1093/ajcn/69.1.30.

6. McKeown NM, Troy LM, Jacques PF, Hoffmann U, O'Donnell CJ, Fox CS. Whole- and refined-grain intakes are differentially associated with abdominal visceral and subcutaneous adiposity in healthy adults: the Framingham Heart Study. *Am J Clin Nutr.* 2010;92(5):1165-71. doi:10.3945/ajcn.2009.29106.

   Kikuchi Y, Nozaki S, Makita M, et al. Effects of Whole Grain Wheat Bread on Visceral Fat Obesity in Japanese Subjects: A Randomized Double-Blind Study. *Plant Foods Hum Nutr.* 2018;73(3):161-165. doi:10.1007/s11130-018-0666-1.

7. Chandler-Laney PC, Morrison SA, Goree LL, Ellis AC, Casazza K, Desmond R, Gower BA. Return of hunger following a relatively high carbohydrate breakfast is associated with earlier recorded glucose peak and nadir. *Appetite.* 2014;80:236-41. doi:10.1016/j.appet.2014.04.031.

8. Spieth LE, Harnish JD, Lenders CM, Raezer LB, Pereira MA, Hangen SJ, Ludwig DS. A low–glycemic index diet in the treatment of pediatric obesity. *Arch Pediatr Adolesc Med.* 2000;154(9):947-951. doi:10.1001/archpedi.154.9.947.

   Ludwig DS, Majzoub JA, Al-Zahrani A, Dallal G, Blanco I, Roberts SB. High glycemic index foods, overeating, and obesity. *Pediatrics.* 1999;103(3):e26; doi:10.1542/peds.103.3.e26.

9. Vanegas SM, Meydani M, Barnett JB, Goldin B, Kane A, Rasmussen H, Brown C, et al. Substituting whole grains for refined grains in a 6-wk randomized trial has a modest effect on gut microbiota and immune and inflammatory markers of healthy adults. *Am J Clin Nutr.* 2017;105(3):635-650. doi:10.3945/ajcn.116.146928.

   Krajmalnik-Brown R, Ilhan ZE, Kang DW, DiBaise JK.. Effects of gut microbes on nutrient absorption and energy regulation. *Nutr Clin Pract.* 2012;27(2):201-14. doi:10.1177/0884533611436116.

10. Chandler-Laney PC, Morrison SA, Goree LL, Ellis AC, Casazza K, Desmond R, Gower BA. Return of hunger following a relatively high carbohydrate breakfast is associated with earlier recorded glucose peak and nadir. *Appetite.* 2014;80:236-41. doi:10.1016/j.appet.2014.04.031.

11. Raben A, Agerholm-Larsen A, Flint A, Holst JJ, Astrup A. Meals with similar energy densities but rich in protein, fat, carbohydrate, or alcohol have different effects on energy expenditure and substrate metabolism but not on appetite and energy intake. *Am J Clin Nutr.* 2003;77(1):91–100. doi:10.1093/ajcn/77.1.91.

12. Barr SB, Wright JC. Postprandial energy expenditure in whole-food and processed-food meals: implications for daily energy expenditure. *Food Nutr Res.* 2010;54. doi:10.3402/fnr.v54i0.5144.

13. Karl JP, Meydani M, Barnett JB, Vanegas SM, Goldin B, Kane A, Rasmussen H, et al. Substituting whole grains for refined grains in a 6-wk randomized trial favorably affects energy-balance metrics in healthy men and postmenopausal women. *Am J Clin Nutr.* 2017;105(3):589-599. doi:10.3945/ajcn.116.139683.

14. Kikuchi Y, Nozaki S, Makita M, et al. Effects of Whole Grain Wheat Bread on Visceral Fat Obesity in Japanese Subjects: A Randomized Double-Blind Study. *Plant Foods Hum Nutr.* 2018;73(3):161-165. doi:10.1007/s11130-018-0666-1.

15. Jonnalagadda SS, et al. Putting the whole grain puzzle together: health benefits associated with whole grains--summary of American Society for Nutrition 2010 Satellite Symposium. *J Nutr.* 2011;141(5):1011S-22S. doi:10.3945/jn.110.132944.

16. Zong G, et al. Whole Grain Intake and Mortality From All Causes, Cardiovascular Disease, and Cancer: A Meta-Analysis of Prospective Cohort Studies. *Circulation.* 2016;133(24):2370-80. doi:10.1161/CIRCULATIONAHA.115.021101.

17. Benincasa P, et al. Sprouted Grains: A Comprehensive Review. *Nutrients.* 2019;11(2):421. doi:10.3390/nu11020421.

    Sprouted Whole Grains. The Whole Grains Council. wholegrainscouncil.org/whole-grains-101/whats-whole-grain-refined-grain/sprouted-whole-grains. Accessed April 27, 2017.

18. Cuadrado C, Hajos G, Burbano C, et al. Effect of Natural Fermentation on the Lectin of Lentils Measured by Immunological Methods. *Food and Agricultural Immunology.* 2002;14(1)41-49. doi:10.1080/09540100220137655.

    Nkhata SG, et al. Fermentation and germination improve nutritional value of cereals and legumes through activation of endogenous enzymes. *Food Science & Nutrition.* 2018;6(8):2446-2458. doi:10.1002/fsn3.846.

    Katina K, Liukkonen KH, et al. Fermentation-induced changes in the nutritional value of native or germinated rye. *Journal of Cereal Science.* 2007;46(3):348-355. doi:10.1016/j.jcs.2007.07.006.

19. Benincasa P, et al. Sprouted Grains: A Comprehensive Review. *Nutrients.* 2019;11(2):421. doi:10.3390/nu11020421.

20. El Khoury D, Cuda C, Luhovyy BL, Anderson GH. Beta glucan: health benefits in obesity and metabolic syndrome. *J Nutr Metab.* 2012;2012:851362. doi:10.1155/2012/851362.

Chapter 11 Munch Mindfully

1. Li J, Zhang N, Hu L, Li Z, Li R, Li C, Wang S. Improvement in chewing activity reduces energy intake in one meal and modulates plasma gut hormone concentrations in obese and lean young Chinese men. *Am J Clin Nutr.* 2011;94(3):709–716. doi:10.3945/ajcn.111.015164.

Miquel-Kergoat S, et al. Effects of Chewing on Appetite, Food Intake and Gut Hormones: A Systematic Review and Meta-Analysis. *Physiol Behav.* 2015;151:88–96. doi:10.1016/j.physbeh.2015.07.017.

Andrade AM, Greene GW, Melanson KJ. Eating slowly led to decreases in energy intake within meals in healthy women. *J Acad Nutr Diet.* 2008;108(7):1186-1191. doi:10.1016/j.jada.2008.04.026.

Zhu B, et al. Association Between Eating Speed and Metabolic Syndrome in a Three-Year Population-Based Cohort Study. *J Epidemiol.* 2015;25(4):332–336. doi:10.2188/jea.je20140131.

2 Shah M, Copeland J, Dart L, Adams-Huet B, James A, Rhea D. Slower Eating Speed Lowers Energy Intake in Normal-Weight but Not Overweight/Obese Subjects. *J Acad Nutr Diet.* 2014;114(3):393-402. doi:10.1016/j.jand.2013.11.002.

3 Tanihara S, et al. Retrospective Longitudinal Study on the Relationship between 8-Year Weight Change and Current Eating Speed. *Appetite.* 2011;57(1)179-183. doi:10.1016/j.appet.2011.04.017.

4 Hamada Y, et al. The Number of Chews and Meal Duration Affect Diet-Induced Thermogenesis and Splanchnic Circulation. *Obesity.* 2014;22(5):E62-9. doi:10.1002/oby.20715.

5 Li J, Zhang N, Hu L, Li Z, Li R, Li C, Wang S. Improvement in chewing activity reduces energy intake in one meal and modulates plasma gut hormone concentrations in obese and lean young Chinese men. *Am J Clin Nutr.* 2011;94(3):709–716. doi:10.3945/ajcn.111.015164.

6 Wansink B, Payne CR, Chandon P. Internal and external cues of meal cessation: The French paradox redux?. *Obesity.* 2007;15(12):2920-2924. doi:10.1038/oby.2007.348.

Chapter 12 Sleep Soundly

1 National Sleep Foundation. 2013 International Bedroom Poll: Summary of Findings. Arlington, VA: National Sleep Foundation; 2013. www.sleepfoundation.org/sites/default/files/inline-files/RPT495a.pdf. Accessed February 26, 2020.

2 Youngstedt SD, Kline CE. Epidemiology of exercise and sleep. *Sleep Biol Rhythms.* 2006;4(3):215-221. doi:10.1111/j.1479-8425.2006.00235.x.

Reid KJ, Baron KG, Lu B, Naylor E, Wolfe L, Zee PC. Aerobic exercise improves self-reported sleep and quality of life in older adults with insomnia. *Sleep Med.* 2010;11(9):934-40. doi:10.1016/j.sleep.2010.04.014.

3 Myllymäki T, Kyröläinen H, Savolainen K, Hokka L, Jakonen R, Juuti T, Martinmäki K, Kaartinen J, Kinnunen ML, Rusko H. Effects of vigorous late-night exercise on sleep quality and cardiac autonomic activity. *J Sleep Res.* 2011;20(1 Pt 2):146-53. doi:10.1111/j.1365-2869.2010.00874.x.

4 Grandner MA. Best of the Rest: Improving Health Through Better Sleep. Presented at FNCE; October 27, 2019; Philidelphia.

5 Mullington JM, Simpson NS, Meier-Ewert HK, Haack M. Sleep loss and inflammation. *Best Pract Res Clin Endocrinol Metab.* 2010;24(5):775-84. doi:10.1016/j.beem.2010.08.014.

6 Morselli LL, Guyon A, Spiegel K. Sleep and metabolic function. *Pflugers Arch.* 2012;463(1):139-160. doi:10.1007/s00424-011-1053-z.

7 Al Khatib HK, Harding SV, Darzi J, Pot GK. The effects of partial sleep deprivation on energy balance: a systematic review and meta-analysis. *Eur J Clin Nutr.* 2017;71(5):614-624. doi:10.1038/ejcn.2016.201.

Lyytikäinen P, Rahkonen O, Lahelma E, Lallukka T. Association of sleep duration with weight and weight gain: a prospective follow-up study. *J Sleep Res.* 2011;20(2):298-302. doi:10.1111/j.1365-2869.2010.00903.x.

8 Kondracki NL. The Link Between Sleep and Weight Gain - Research Shows Poor Sleep Quality Raises Obesity and Chronic Disease Risk. *Today's Dietitian.* 2012;14(6):48. www.todaysdietitian.com/newarchives/060112p48.shtml.

9 Grandner MA. Best of the Rest: Improving Health Through Better Sleep. Presented at FNCE; October 27, 2019; Philidelphia.

10 Krause AJ, Simon EB, Mander BA, Greer SM, Saletin JM, Goldstein-Piekarski AN, Walker MP. The sleep-deprived human brain. *Nat Rev Neurosci.* 2017;18(7):404-418. doi: 10.1038/nrn.2017.55.

11 Payne JD, Nadel L. Sleep, dreams, and memory consolidation: the role of the stress hormone cortisol. *Learn Mem.* 2004;11(6):671-8. doi:10.1101/lm.77104.

12 Centers for Disease Control and Prevention. Drowsy Driving. Centers for Disease Control and Prevention. www.cdc.gov/sleep/about_sleep/drowsy_driving.html. Updated March 21, 2017. Accessed February 24, 2020.

13 Buchmann N, Spira D, Norman K, Demuth I, Eckardt R, Steinhagen-Thiessen E. Sleep, Muscle Mass and Muscle Function in Older People. *Dtsch Arztebl Int.* 2016;113(15):253-60. doi:10.3238/arztebl.2016.0253.

14 Hairston KG, Bryer-Ash M, Norris JM, Haffner S, Bowden DW, Wagenknecht LE. Sleep duration and five-year abdominal fat accumulation in a minority cohort: the IRAS family study. *Sleep.* 2010;33(3):289-95. doi:10.1093/sleep/33.3.289.

15 Chaput JP, Després JP, Bouchard C, Tremblay A. The association between sleep duration and weight gain in adults: a 6-year prospective study from the Quebec Family Study. *Sleep.* 2008;31(4):517-23. doi:10.1093/sleep/31.4.517.

16 Nedeltcheva AV, Kilkus JM, Imperial J, Schoeller DA, Penev PD. Insufficient sleep undermines dietary efforts to reduce adiposity. *Ann Intern Med.* 2010;153(7):435-41. doi:10.7326/0003-4819-153-7-201010050-00006.

17 Morris CJ, Fullick S, Gregson W, Clarke N, Doran D, MacLaren D, Atkinson G. Paradoxical post-exercise responses of acylated ghrelin and leptin during a simulated night shift. *Chronobiol Int.* 2010;27(3):590-605. doi:10.3109/07420521003663819.

Calvin AD, Carter RE, Levine JA, Somers VK. Abstract MP030: Insufficient Sleep Increases Caloric Intake but not Energy Expenditure. *Circulation.* 2012;125(suppl_10):AMP030. doi:10.1161/circ.125.suppl_10.amp030.

18 King's College London. Sleep deprivation may cause people to eat more calories. ScienceDaily. www.sciencedaily.com/releases/2016/11/161102130724.htm. Published November 2, 2016. Accessed February 21, 2016.

19. Stengel A1, Taché Y. Ghrelin–a pleiotropic hormone secreted from endocrine X/A-like cells of the stomach. *Front Neurosci*. 2012;6:24. doi:10.3389/fnins.2012.00024.

20. Nedeltcheva AV, Kilkus JM, Imperial J, Schoeller DA, Penev PD. Insufficient sleep undermines dietary efforts to reduce adiposity. *Ann Intern Med*. 2010;153(7):435-41. doi:10.7326/0003-4819-153-7-201010050-00006.

21. Shechter A, Grandner MA, St-Onge MP. The Role of Sleep in the Control of Food Intake. *Am J Lifestyle Med*. 2014;8(6):371-374. doi:10.1177/1559827614545315.

    St-Onge MP. The role of sleep duration in the regulation of energy balance: effects on energy intakes and expenditure. *J Clin Sleep Med*. 2013;9(1):73-80. doi:10.5664/jcsm.2348.

22. Greer SM, Goldstein AN, Walker MP. The impact of sleep deprivation on food desire in the human brain. *Nat Commun*. 2013;4:2259. doi:10.1038/ncomms3259.

23. Crispim CA, Zimberg IZ, dos Reis BG, Diniz RM, Tufik S, de Mello MT. Relationship between food intake and sleep pattern in healthy individuals. *J Clin Sleep Med*. 2011;7(6):659-64. doi: 10.5664/jcsm.1476.

24. St-Onge MP, Roberts A, Shechter A, Choudhury AR. Fiber and saturated fat are associated with sleep arousals and slow wave sleep. *J Clin Sleep Med*. 2016;12(1):19-24. doi:10.5664/jcsm.5384.

25. Krueger JM. The role of cytokines in sleep regulation. *Curr Pharm Des*. 2008;14(32):3408-16. doi:10.2174/138161208786549281.

26. Prather AA, Janicki-Deverts D, Hall MH, Cohen S. Behaviorally Assessed Sleep and Susceptibility to the Common Cold. *Sleep*. 2015;38(9):1353-9. doi:10.5665/sleep.4968.

    Besedovsky L, Born J. Sleep, Don't Sneeze: Longer Sleep Reduces the Risk of Catching a Cold. *Sleep*. 2015;38(9):1341-2. doi:10.5665/sleep.4958.

27. St-Onge MP, Roberts A, Shechter A, Choudhury AR. Fiber and saturated fat are associated with sleep arousals and slow wave sleep. *J Clin Sleep Med*. 2016;12(1):19-24. doi:10.5664/jcsm.5384.

28. Santa Cruz J. The Link Between ZZZs & Eats. *Today's Dietitian*. 2019;32(21)8:32. https://www.todaysdietitian.com/newarchives/0819p32.shtml.

29. Grandner MA, Jackson N, Gerstner JR, Knutson KL. Dietary nutrients associated with short and long sleep duration. Data from a nationally representative sample. *Appetite*. 2013;64:71-80. doi:10.1016/j.appet.2013.01.004.

30. Grandner MA, Jackson N, Gerstner JR, Knutson KL. Dietary nutrients associated with short and long sleep duration. Data from a nationally representative sample. *Appetite*. 2013;64:71-80. doi:10.1016/j.appet.2013.01.004.

31. Kant AK, Graubard BI. Association of self-reported sleep duration with eating behaviors of American adults: NHANES 2005-2010. *Am J Clin Nutr*. 2014;100(3):938-947. doi:10.3945/ajcn.114.085191.

32. Grandner MA, Jackson N, Gerstner JR, Knutson KL. Dietary nutrients associated with short and long sleep duration. Data from a nationally representative sample. *Appetite*. 2013;64:71-80. doi:10.1016/j.appet.2013.01.004.

33. Haghighatdoost F, Karimi G, Esmaillzadeh A, Azadbakht L. Sleep deprivation is associated with lower diet quality indices and higher rate of general and central obesity among young female students in Iran. *Nutrition*. 2012;28(11-12):1146-1150. doi:10.1016/j.nut.2012.04.015.

34. Tanaka E, Yatsuya H, Uemura M, et al. Associations of protein, fat, and carbohydrate intakes with insomnia symptoms among middle-aged Japanese workers. *J Epidemiol*. 2013;23(2):132-138. doi:10.2188/jea.JE20120101.

35. Tanaka E, Yatsuya H, Uemura M, et al. Associations of protein, fat, and carbohydrate intakes with insomnia symptoms among middle-aged Japanese workers. *J Epidemiol*. 2013;23(2):132-138. doi:10.2188/jea.JE20120101.

36. Castro-Diehl C, Wood AC, Redline S, et al. Mediterranean diet pattern and sleep duration and insomnia symptoms in the Multi-Ethnic Study of Atherosclerosis. *Sleep*. 2018;41(11):zsy158. doi:10.1093/sleep/zsy158.

    Campanini MZ, Guallar-Castillón P, Rodríguez-Artalejo F, Lopez-Garcia E. Mediterranean diet and changes in sleep duration and indicators of sleep quality in older adults. *Sleep*. 2017;40(3). doi:10.1093/sleep/zsw083.

37. Santa Cruz J. The Link Between ZZZs & Eats. *Today's Dietitian*. 2019;32(21)8:32. https://www.todaysdietitian.com/newarchives/0819p32.shtml.

38. Watson EJ, Coates AM, Kohler M, Banks S. Caffeine Consumption and Sleep Quality in Australian Adults. *Nutrients*. 2016;8(8):479. doi:10.3390/nu8080479.

    Chaudhary NS, Grandner MA, Jackson NJ, Chakravorty S. Caffeine consumption, insomnia, and sleep duration: Results from a nationally representative sample. *Nutrition*. 2016;32(11-12):1193-9. doi:10.1016/j.nut.2016.04.005.

39. Florida Atlantic University. Sleep interrupted: What's keeping us up at night?. ScienceDaily. Published 6 August 2019. www.sciencedaily.com/releases/2019/08/190806101604.htm.

40. Grandner MA. Best of the Rest: Improving Health Through Better Sleep. Presented at FNCE; October 27, 2019; Philidelphia.

CONTRIBUTORS

Judith (aka Judes) Scharman Draughon MS, RDN, LDN is a registered, licensed dietitian nutritionist, the author of the book, *Lean Body Smart Life* and the wellness plan, *12 Fixes to Healthy*. She inspires many groups, organizations, and businesses with her high-energy nutrition presentations, workshops, and seminars.

Judes narrows down the science to the most effective ways to improve your health and weight for life and has been dubbed the "How-To-Dietitian" as she shows how to prepare deliciously health-promoting foods while juggling busy lives. She is the owner of Nutrition Educational Solutions, but the world knows her as "Foods With Judes." Besides Judes' corporate wellness work, she taught at the International Culinary Arts and Sciences Institute and held various other positions as a dietitian through the years. She is the mother of four adult children and resides in North Carolina with her husband.

LinkedIn: Nutrition Educational Solutions (Foods with Judes)
Instagram: @foodswithjudes, Facebook: @foodswithjudes, Twitter: @foodswithjudes

Traci Fisher began learning about health and leadership in the United States military and as a helicopter pilot in the US Army. She understands how important it is to be healthy in both mind and body. She has been helping leaders align their health values with their leadership values for over fifteen years. Traci is certified through the National Association of Sports Medicine, Wellcoaches and the Titleist Performance Institute. She is the managing director of Total Customized Fitness, creator of The Wellness Coach, featured in Live Active Heart Healthy Workouts Fitness Videos and is a nationally certified health & wellness coach. Traci is also a contributing author to *Lean Body, Smart Life* and *12 Fixes to Healthy* and is looking forward to supporting you on your wellness journey.

Natalie Stephens is a Registered Dietitian and Life Coach. She is also the founder of Dieting Differently. Natalie specializes in helping clients manage their appetite and cravings. She teaches clients how to self-regulate food intake and crave healthier living. Her focus is on shifting perspective, creating motivation, and using micro-habits to build new routines that make nutrition and wellness an intuitive part of your life. Natalie believes eating healthy shouldn't consume your life, it should enrich your life.

For the last eight years, Natalie has been showing people with acute and chronic conditions how to use nutrition to better their lives. She has worked as a dietitian with people struggling with disordered eating, digestive disorders, diabetes, heart disease, PCOS, kidney failure (dialysis), and liver disease. Natalie resides in the DC area with her husband and three daughters and loves high adventure.

INDEX

A

added sugar
   in energy drinks, 127
   identifying on food labels, 20–22
   in kefir, 131
   reducing in diet, 26–27
   in soda and sweet beverages, 18
   in yogurt, 59
ADHD, 105
agave nectar, 17, 19
age-related muscle loss, 29–30
ALA, 164, 167, 169, 170, 177. *See also* omega-3 fatty acids
alcohol
   consumption before sleeping, 90, 233
   in kefir, 52, 130
   in kombucha, 128–129
   mitigating the effects of, 131–134
   physical activity and reduction of cravings, 105
   physical effects of, 131
Allen, Linda, 212
allergies, 16, 73, 167
Alzheimer's disease, 104, 105, 166, 225
amaranth, 22, 24, 30, 195, 197, 203
American Journal of Clinical Nutrition, 192
American Journal of Nutrition, 211
amylase-trypsin inhibitors, 23
ancient grains, 22, 24, 197. *See also* whole grains; *specific grains* ~
antioxidants
   in dark chocolate and cocoa, 17
   in dates, 16
   in eggs from foraging hens, 38
   in fats, 182
   in produce, 86
   and selenium, 167
   in sprouted grains, 195
Anton, Steve, 98
anxiety, 62, 127, 178
arthritis, 46, 165
artificial sweeteners, 17–18
artisan bread, 201
asthma, 167
attention span, 31–33, 105
autism, 46
autoimmune diseases, 23, 46, 48
avocado oil, 170, 174, 183, 187
avocados
   cooking with, 181
   and electrolytes, 134, 135
   fiber in, 51
   as a monounsaturated fat, 170
   and prebiotics, 55
   replacing unhealthy fats with, 172, 173, 174

B

baking with whole foods, 14, 204
batch cooking, 146–147
beans, 70, 75–76
belly fat
    fructose and, 18, 19, 20
    impact of whole grains on, 190, 193
    reduction of with omega-9 fatty acids, 170
    reduction of with sleep, 227
    and saturated fats, 170
beta-amyloid, 225
beta-carotene, 38, 66
Bifidobacterium, 48
biotin, 37, 61
black rice, 8, 24, 191
blood pressure, 40, 55, 190
blood sugar. *See* glucose
body fat
    effect of prebiotics on, 55
    and effects of eating produce, 66
    effects of intermittent fasting on, 92, 98
    effects of nighttime eating on, 91
breakfast
    advantages of eating, 91, 93
    and the circadian rhythm, 89–91, 94, 100
    and intermittent fasting, 91–92, 98
    plan for eating, 101–102
    protein with, 92
    and weight loss, 90–91, 92, 93, 100–101
brown rice, 8, 24, 51, 56
brown sugar, 19. *See also* fructose; *See also* sugar
buckwheat, 24, 35, 195
butter, 172–173

C

caffeine
    effects of, 127
    in energy drinks, 127
    in kefir, 52, 130
    in kombucha, 129
    and sleep quality, 127, 227
    and switching from soda to water, 124–126
cage-free eggs labeling, 37–38, 177. *See also* food labels
calcium
    benefits of, 117
    in cottage cheese, 54
    in dates, 15
    deficiency in vegans, 37
    effects of deficiency in, 96
    and electrolytes, 135
    in kefir, 131
    in yogurt, 40, 57, 58, 62
caloric beverages, 122–123. *See also* sodas
caloric intake, 3, 12–13
calorie consumption, 46–47, 90–91, 151
cancer
    effects of omega-3 fatty acids on, 165
    effects of produce on, 66
    effects of whole foods on, 12
    and imbalance of microbiome, 46
    impact of whole grains on, 190
    reducing chance of with physical activity, 105
    role of whole grains in reduction of, 22
canola oil, 170–171, 183
carbohydrates
    and exercise, 113–117, 196
    and FODMAP, 24
    and whole grains, 22, 190, 208
carbonated water, 125–126. *See also* water
cardiovascular disease, 40, 46, 54, 105, 178
celiac disease, 23, 46, 48
cheese, 55
chewing food, 209–212, 223–224
chia seeds, 24, 167, 182, 184
childhood allergies, 167. *See also* allergies
chocolate, dark. *See* dark chocolate
chocolate milk, 116
cholesterol, 37, 38, 170
choline, 33, 37, 169
cinnamon, 15
circadian rhythm
    definition of, 89
    eating according to, 89–91, 94, 100–101
    plan for eating according to, 101–102
circulating triglycerides, 40, 55
club soda, 125, 138
cocoa, 14, 17
coconut oil, 173–174
coconut water, 135
cognitive function
    breakfast and, 31–33
    effects of exercise on, 104
    effects of omega-3 fatty acids on, 165–166, 178
    and gut health, 48, 62
    importance of produce in, 67, 87
    importance of quality sleep for, 227

physical activity on teenagers', 107
colon health, 50
complex carbohydrates, 196. *See also* carbohydrates
cooking with whole foods, 75–85
corn, 22
cortisol, 105
cost considerations of whole foods, 155–157
cottage cheese, 41, 54, 62
couscous, 203
cravings. *See* food cravings
cultured dairy, 39, 49, 55. *See also* fermented foods
cultured foods. *See* fermented foods

D

dairy
    benefits of consuming, 39, 54–55, 95
    and carbohydrates, 114, 116
    and dieting, 24, 25
    before exercise, 196
    fermented, 39, 40, 55, 62, 131
    and protein, 36, 39
    protein contained in, 35
    and saturated fats, 164, 173
    sensitivities, 40, 57–58
dark chocolate, 14, 17, 231
dates, 15
dehydration
    and alcohol, 133
    associations with caffeine, 127
    and cognitive abilities, 123
    and headaches, 127
    prevention of while exercising, 114–117, 134–136
dementia, 166
depression
    effects of omega-3 fatty acids on, 166, 178
    food's role in, 13
    and imbalance of microbiome, 46
    reduction of with physical activity, 105
    and relationship to processed carbohydrates, 190
    relation to gut health, 62
DHA, 164, 165, 166, 167, 169, 177. *See also* omega-3 fatty acids
Diabetes, Obesity and Metabolism, 124
Dietary Inflammatory Index (DII), 86
diets
    and common associations, 218–220
    Keto, 47, 95–98
    low-FODMAP, 24–25
    Mediterranean, 4, 47, 231–232
diet sodas, 18, 124, 126. *See also* sodas
digestion, 48, 211–212. *See also* gut health
diseases, 12, 13, 66
dopamine, 215

E

early time-restricted feeding (eTRF), 91–92, 93. *See also* intermittent fasting
eating in restaurants, 152–154, 160
edamame, 77–78
eggs
    omega-3 fatty acids in, 169, 177
    as a protein source, 37
    selection of healthiest, 37–38, 176–177
einkorn, 22
electrolytes, 134–136
energy drinks, 127
EPA, 164, 169, 177. *See also* omega-3 fatty acids
exercise. *See also* movement; *See also* physical activity
    and brain development of unborn babies, 110
    effect on weight loss, 3
    and electrolytes, 134–136
    and nutrition for energy, 113–117
    preventing dehydration during, 114–117

F

farro, 22, 203
fasting. *See* intermittent fasting
fats. *See also* specific fats
    cooking with, 181–183
    kinds of, 163–164
    and nighttime eating, 229
    replacing unhealthy with healthy, 186–187
    shopping for, 174–179
fermented foods
    benefits of, 61
    cultured dairy, 39
    and gut health, 53
    importance of, 49
    kombucha, 128–130
    process of making, 53
    types of, 54–55, 62
    whole grains, 195, 208
fiber

amount in common whole foods, 51, 191
amount to consume, 190–191
benefits of, 190–193
and effect on sleep, 230
effects on the body, 16
and gut health, 50, 128
and prebiotics, 56
in whole grains, 189, 190
fish. *See* seafood
Fisher, Traci, 104
fish oil supplements, 165
FITT principle, 110, 118–119
flavonoids, 17
flavored water, 124–125, 137–138
flour, whole grain, 199–200
FODMAP, 24–25, 57
folate, 15, 177, 195, 197, 230
food allergies. *See* allergies
food cravings
    and alcohol, 133
    controlling, 220–222
    effects of whole foods on, 12
    and intermittent fasting, 98
    protein and, 29
    sugar's role in, 14
food labels
    and added sugar, 20–21
    and caffeine, 127
    on eggs, 37, 177
    and GMOs, 74
    on grain products, 197–198, 200
    identifying processed foods on, 71–72
    on seafood, 178
    and serving sizes, 21–22
food sensitivities, 23. *See also* allergies
food shaming, 96
freekeh, 22, 197, 203
free-range eggs labeling, 37–38
fructans, 23
fructose, 19, 20

G

genetically modified organisms (GMOs), 73–74
genetics, 48
glucose
    and artificial sweeteners, 18
    and breakfast, 33, 101
    and circadian rhythm, 89–90
    and fruits, 19
    and high-fructose corn syrup, 19

and keto diet, 95, 98
role in body fat, 20
in soda and sweet beverages, 20
and whole grains, 190, 206
glucose levels
    and cinnamon, 15
    controlling with fermented foods, 53
    controlling with omega-9 fatty acids, 170
    controlling with physical activity, 108
    and diet sodas, 124
    effects of overnight fasting on, 93
    effects of prebiotics on tolerance of, 55
    effects of processed foods on, 114
    effects of trans fats on, 172
    and oats, 201
    and resistant starches, 202
    yogurt consumption on control of, 40, 55
gluten, 23
gluten-free grains, 23–24, 195
gluten-free processed foods, 23, 49
glycemic index (GI), 16
goat milk yogurt, 58
gout, 46
grain bowls, 204–205
Grandner, Michael, 227
Greek yogurt, 39, 57
growth hormone, 90, 101
gut health
    and diet sodas, 18
    effects of caffeine on, 127
    effects of nighttime eating on, 91
    effects of processed foods on, 23
    effects of produce on, 66–67
    and importance of breastfeeding, 50
    and importance of microbiota, 45
    plan for improving, 63
    processed food's effect on, 13
    and whole grains, 191–192

H

habits and routines, 139–143, 158–159, 233, 235–236
Harrell, Allison, 147
HDL cholesterol, 170, 172
heart attacks, 93
heart disease
    due to trans fats, 172
    effects of omega-3 fatty acids on, 165
    effects of whole foods on, 12

impact of whole grains on, 190
and saturated fat, 173
sugar's role in, 14
high blood pressure. *See* blood pressure
high-calorie foods, 3
high-fructose corn syrup, 17–18, 19–20
honey, raw. *See* raw honey
hormone imbalances, 228
hunger cravings, 98. *See also* food cravings
hunger signals, 212–217, 223–224
hydration, 113–114, 121–122, 123. *See also* dehydration; *See also* water
hypertension, 46

I

Icelandic yogurt, 39, 57
immune system. *See also* gut health
effects of quality sleep on, 229
effects of whole foods on, 12
importance of microbiota on, 45
processed food's effect on, 13
whole grains' support of, 22
infant health, 46
inflammation
effects of omega-3 fatty acids on, 165, 178
effects of produce on, 66
effects of whole foods on, 12
fructose and, 18
and imbalance of microbiome, 46
processed food's effect on, 13
and quality sleep, 227
infused water, 125, 137–138
insulin, 18, 40, 89–90, 172
insulin resistance, 55, 227
intermittent fasting
benefits of, 93
and circadian rhythm, 91–92
flexibility with, 99
versus keto diet, 95–98
and weight loss, 92–93, 100–101
International Journal of Obesity, 90–91, 100
iodine, 169
iron, 15, 37, 183, 190, 195
irritable bowel syndrome (IBS), 24, 52, 165

J

Journal of Clinical Endocrinology and Metabolism, 123
Journal of Clinical Sleep Medicine, 230
Journal of Nutrition, 30

Journal of the International Society of Sports Nutrition, 31

K

Kalman, Douglas, 124
kamut, 22, 203
Katz, David, 173
kefir
benefits of, 130–131
non-dairy options, 62–63
as a probiotic food, 40, 54
role in gut health, 52
substituting with, 60
keto diet, 47, 95–98
kimchi, 53
kinetic chain, 106
kombu, 76, 77
kombucha, 52, 128–130, 130

L

lactic acid, 53, 61, 195
Lactobacillus, 48
lactose, 54
lactose intolerance, 40
LDL cholesterol, 170, 172
lectin, 77
legumes, 55, 63, 70, 75
Leidy, Heather, 31–33
lentils, 70, 75
leucine, 39, 54
live active bacteria
benefits of, 45
in cheese, 41
in fermented foods, 53–54, 61, 62–63
in probiotic supplements, 52
longevity, 46, 68, 193
lutein, 38

M

maple syrup, pure. *See* pure maple syrup
Mealime, 146
meal planning, 146–148, 159–161
Mediterranean diet, 4, 47, 231–232
medium-chain fatty acids, 58, 173
melatonin, 90, 101, 230
memory and memory loss, 31–33, 166. *See also* cognitive function
mental health. *See* mood and mood disorders
mercury toxicity, 39, 167, 168, 178

metabolism, 100–101, 192–193, 211, 228–229
microbes. *See* live active bacteria
microbiomes and microbiota
    benefits of, 45–48
    definition of, 45
    and diet sodas, 18
    effect of breastfeeding on, 50
    effects of processed foods on, 23
    maintaining with whole foods, 23, 128
    and prebiotics, 55–56
    and probiotic supplements, 52
    role of diet in balancing, 48–49, 61–62
    and whole grains, 191–192
milk-based kefir, 130–131. *See also* kefir
millet, 22, 24
mindful eating
    and chewing food, 209–212, 223
    and diet associations, 218–220
    and food cravings, 220–222
    and hunger signals, 212–217
miso, 53, 62, 63, 78
monk fruit, 18
monounsaturated fatty acids, 170–171, 173
mood and mood disorders
    associations with caffeine, 127
    effect of prebiotics on, 55
    effects of omega-3 fatty acids on, 166–167, 178
    and gut health, 48
    and imbalance of microbiome, 46
    and physical activity, 105
    and relationship to processed carbohydrates, 190
    relation to gut health, 62
morning eating. *See* breakfast
movement
    adaptation by the body to, 109
    and alcohol, 134
    benefits of, 105–106, 108
    definition of, 105–106
    effects on cognitive function in teenagers, 107
    improving quality of, 108–109, 111, 118–119
    increasing intensity of, 112
    increasing quantity of, 111, 118–119
    plan for moving more, 119–120
    and proper nutrition, 113–114
    and quality sleep, 225–226
    recommended levels of, 110
    types of, 110
movement spectrum, 107
Mozaffarian, Darius, 39
multiple sclerosis, 46
muscle loss, 29–30

N

National Institutes of Health, 12, 68
natto, 78
New England Journal of Medicine, 3, 47
nighttime eating, 90, 101, 229, 235–236
non-dairy foods, 39, 58, 62–63
non-plant foods, 36–37
nut butters, 3, 175
nuts and seeds
    effect on weight loss, 3
    and omega-9 fatty acids, 170
    and prebiotics, 55, 63
    shopping for, 176
    ways to prepare, 182–184

O

oats and oatmeal, 24, 195, 201, 202
obesity
    and imbalance of microbiome, 46, 61
    and quality sleep, 227
    and relationship to processed carbohydrates, 190
    role of whole grains in reduction of, 22
    soda's role in, 18
    sugar's role in, 14
olive oil, 171, 181
omega-3 fatty acids
    benefits of, 163–164, 178
    content in seafood products, 168
    in eggs, 38, 169, 177
    types of, 164–165
omega-9 fatty acids, 170–171. *See also* monounsaturated fatty acids
organic foods, 72–73
osteoporosis, 105

P

pancake syrup, 17
Panda, Satchidananda, 101
pasteurization, 54
pasture-raised eggs labeling, 38, 176–177. *See also* food labels
peanut butter, 174–175
peanuts, 176
phosphorus, 40, 54
physical activity, 105–106, 108. *See*

also exercise; See also movement
phytonutrients
    benefits of, 67
    in extra virgin olive oil, 171
    number in plant-based whole foods, 11
    in produce, 86
    in soy, 77
    in whole grains, 22
plant-based omega-3 fat foods, 167
plant-based whole foods, 36–37. See also whole foods; See also whole grains
polyunsaturated fats, 164–169, 173
popcorn, 24
portion control, 144–145, 151–154, 159
potassium, 40, 54, 96, 116, 134, 183
potatoes. See specific potatoes
prebiotics, 55–56, 63
pregnancy, 110, 165, 166, 167
preparing whole foods, 75–85
probiotic foods, 54–55
probiotics, 40, 49, 50, 52, 53, 58
probiotic supplements, 52
processed foods
    effects on gut health, 63
    how to identify, 12
    replacing with whole foods, 26–27, 182–183, 196
    role in weight gain, 12–13
    and sugar, 26
    and trans fats, 172
processed sugars, 19, 26–27. See also artificial sweeteners
produce
    benefits of eating, 65–68, 86–87
    effect on weight loss, 3
    fruit as a whole food sweetener, 11, 14
    fruits linked to better sleep, 230
    and gut health, 49–50
    increasing consumption of, 142–144, 146–151
    phytonutrients in, 67
    plan for consuming more, 87–88
    and prebiotics, 55–56, 63
    replacing snacks with, 70
    shopping for, 69–71
    water content in, 122
    ways to cook, 75–85
protein
    age-related muscle loss and, 29–30
    benefits of eating enough, 29–33
    breakfast high in, 31–33, 43, 92, 101
    and exercise, 36, 114–117
    kidney health and, 31
    plan for consuming daily need, 42–44
    recommended amounts, 33–36
    role in building muscle, 30
    and sleep patterns, 230
    sources of, 32, 35, 36–41, 43
protein supplements, 40, 43, 114, 117, 200
pulses, 35, 70, 77, 156
pure maple syrup, 6, 11, 14, 17, 26, 27. See also sweeteners, whole food
purple potatoes, 27, 56, 79, 87

Q

quinoa, 22, 24, 195, 203

R

Ranks, Ruth, 125
raw honey, 11, 14, 16. See also sweeteners, whole food
recommended daily allowance (RDA), 34
red rice, 56, 202
refined flours, 22, 26–27, 57
refined honey, 19
refined sugars, 26–27. See also artificial sweeteners; See also processed sugars
resistant starches, 56, 202
restaurant eating, 151–154, 160
rewarding behavior, 140
rheumatoid arthritis, 46
rice. See specific rices
routines and habits. See habits and routines

S

salads and salad dressings, 82–84, 160, 186, 187
saturated fats, 38, 172–173
seafood
    benefits of eating, 39, 165–169, 178
    farmed, 179, 180
    purchasing healthiest, 178–181
    recommended servings of, 166, 178
    and sustainability, 178, 180
    ways to cook, 184–185
Seafood Watch, 181
selenium, 39, 167–169, 178
seltzer water, 125–126, 137. See also water
serotonin, 230
serving sizes, 21. See also portion control
sheep's milk yogurt, 58
short-chain fatty acids, 58

Simopoulos, Artemis, 169
skyr yogurt, 57
sleep
　action plan for sleeping soundly, 237
　benefits of, 225, 226–230
　changing habits for better quality, 233–234, 235–236
　and dark chocolate, 231
　effect on weight loss, 3
　effects of caffeine on, 127
　and exercise, 225–226
　and fiber intake, 230
　fruits that contribute to healthier, 230
　and a Mediterranean diet, 231
　and problems with nighttime eating, 90, 229, 235–236
　and protein intake, 231
　and sugar intake, 231
snacking with whole foods, 148–151, 154–155, 160, 180, 182–184
sodas, 18, 20, 122–123. *See also* diet sodas
soda water, 125–126. *See also* water
sorghum, 22, 24, 195, 203
sourdough bread, 53, 195, 199
soy, 78–79
sparkling water, 125–126. *See also* water
spelt, 22, 203
sports drinks, 135–136
sprouted bread, 198
sprouted grains, 195, 197–198, 208
Stephens, Natalie, 203
Stookey, Jodi, 122–123
stroke, 93, 105, 165, 172, 178
sucrose, 20
sugar
　effects on gut health, 57, 63
　effects on the body, 13–14
　in energy drinks, 127
　in kefir, 131
　in kombucha, 129
　and sleep patterns, 231
supplements
　fish oil, 165
　probiotic, 52
　protein, 40, 43, 114, 117, 200
　versus whole foods, 36, 66, 164
sweeteners, whole food, 11, 14–17
sweet potatoes, 27, 56, 78, 80, 87, 135

T

table syrup, 17

teff, 22, 24, 197
tempeh, 53, 78
trans fats, 172
triglycerides, 93, 165, 170
tryptophan, 131, 231
type 2 diabetes
　and diet sodas, 18, 124
　due to trans fats, 172
　effects of whole foods on, 12
　and imbalance of microbiome, 46
　impact of whole grains on, 190
　reducing chance of with physical activity, 105
　role of whole grains in reduction of, 22
　soda's role in, 18
　sugar's role in, 14
　and yogurt consumption, 40, 54

V

vegan diets, 36–37
vegetables and fruits. *See* produce
vegetarian-fed eggs labeling, 38
veggie straws and chips, 71
vitamin A, 38, 67, 87, 183
vitamin B12, 37, 40, 54, 61, 177
vitamin C, 67, 87, 167, 195, 197
vitamin D
　in eggs, 37, 38
　in kefir, 131
　in seafood, 169
　in yogurt, 40, 54
vitamin E, 38, 67, 87, 177
vitamin K, 61

W

water
　amount to consume, 124, 137
　benefits of drinking water, 121–123
　content in produce, 122
　dehydration and cognitive abilities, 123
　and dehydration headaches, 127
　effect on metabolism, 123
　and electrolytes, 134–136
　versus energy drinks, 127
　flavoring and infusing suggestions, 124–125, 137–138
　plan for consuming more, 138
　replacing soda with, 124, 126
　role in fat burning, 121–122, 123
　types of sparkling water, 125–126
water kefir, 130. *See also* kefir

weight control, 61, 66
weight loss
    changing routines, 141
    and chewing slowly, 210–211
    effects of produce on, 68
    effects of whole foods on, 12, 190, 192–193
    with intermittent fasting, 91–93
    and quality sleep, 226, 227, 228–229
    skipping meals as a means of, 94
    and timing of meals, 90–91, 100–101
    and water consumption, 123
    and yogurt consumption, 40
wheat gluten, 23
whey protein, 39, 54
white sugar, 19. *See also* fructose
Whole30 diet, 25
whole foods
    benefits of, 11
    budgeting for, 155–156
    cooking fats with, 181–185
    and gut health, 48–50
    increasing consumption of, 146–151
    meal planning with, 146–148, 159–161
    planning snacks with, 148–151, 154–155
    replacing with, 186–187
    and restaurants, 152–155, 160
    substituting with, 153–155, 172–173, 182
    ways to prepare, 75–80
whole food sweeteners. *See* sweeteners, whole food
whole grains. *See also* ancient grains
    amount to consume, 190–191, 193–194, 207
    benefits of, 22, 189–193, 206
    definition of, 190
    effect on weight loss, 3, 192–193
    fermented, 195
    gluten-free choices, 24, 195
    and gut health, 56
    plan for consuming more, 208
    and prebiotics, 55, 63
    preparation of, 201–205
    selection of, 197–200
    sprouted, 195, 197–198, 208
    types of flour, 199–200
wild rice, 24, 202, 203
Willett, Walter, 173
working out. *See* exercise
worry, 234

Y

yogurt
    amount of protein in, 35
    benefits of, 40, 48, 54, 62
    effect on weight loss, 3
    making your own, 59
    non-dairy options, 58, 62–63
    as a probiotic food, 54

ADDITIONAL RESOURCES

WEBSITES

12 Fixes to Healthy 12FixesToHealthy.com
Foods With Judes FoodsWithJudes.com
Eat Right Academy of Nutrition and Dietetics Eatright.org
EatingWell www.eatingwell.com
Seafood Watch www.seafoodwatch.org
Seafood Nutrition Partnership www.seafoodnutrition.org
Wild Alaska Seafood www.wildalaskaseafood.com
Norwegian Seafood seafoodfromnorway.us/norwegian-seafood
Vital Choice Wild Seafood & Organics www.vitalchoice.com
Mindful Eating thecenterformindfuleating.org
The Wellness Coach www.thewellness.coach
Dieting Differently DietingDifferently.com

APPS

Hydration Plant Nanny, My Water Balance, Daily Water
Mindful Eating In the moment, Am I Hungry, Slow Eat, Mindful Bite
Sleep Insight Timer

Buying Sustainable Seafood Seafood Watch
12 Fixes To Healthy A Wellness Plan for Life
Mealime Meal Planning App for Healthy Eating

PODCASTS

Liz's Healthy Table with Liz Weiss, MS, RDN (www.lizshealthytable.com/podcast)

Food Psych Podcast with Christy Harrison, MPH, RD, CDN (christyharrison.com/foodpsych)

Sound Bites with Melissa Joy Dobbins, MS, RDN, CDE (www.soundbitesrd.com/podcast)

Body Kindness with Rebecca Scritchfield, MA, RDN (www.bodykindnessbook.com/podcast)

Food Heaven with Wendy Lopez, RDN and Jessica Jones, RDN (foodheaven-madeeasy.com/podcast)

The Less Stressed Life with Christa Biegler, RD, CLT (lessstressedlife.libsyn.com)

Mary's Nutrition Show with Mary Purdy, RD (marypurdy.co/podcast-posts)

Nutrition Diva's Quick and Dirty Tips for Eating Well and Feeling Fabulous (podcasts.apple.com/us/podcast/nutrition-divas-quick-dirty-tips-for-eating-well-feeling)

ARTICLES

The Microbiome and Celiac Disease: A Bacterial Connection by Amy Burkhart MD, RD (theceliacmd.com/articles/microbiome-celiac-disease-bacterial-connection)

BOOKS

12 Fixes to Healthy Workbook by Judith Scharman Draughon, MS, RDN, LDN

Meals that Heal by Carolyn Williams, PhD, RD

The Best 3-Ingredient Cookbook by Toby Amidor, MS, RD, CDN

The Healthy Meal Prep Cookbook by Toby Amidor, MS, RD, CDN

The Easy 5-Ingredient Healthy Cookbook by Toby Amidor, MS, RD, CDN

The Greek Yogurt Kitchen by Toby Amidor, MS, RD, CDN

The Mediterranean DASH Diet Cookbook by Abbie Gellman, MS, RD, CDN

Sixth Edition of Nancy Clark's Sports Nutrition Guidebook by Nancy Clark, MS, RD

Prediabetes: A Complete Guide by Jill Weisenberger, MS, RDN, CDE, CHWC, FAND

Body Kindness by Rebecca Scritchfield, RDN

The 30-Minute Mediterranean Diet Cookbook By Serena Ball, RD & Deanna Segrave-Daly, RD

Does This Clutter Make My Butt Look Fat? By Peter Walsh

The Mindful Eating Workbook By Vincci Tsui, RD

The Mediterranean Way of Eating: Evidence for Chronic Disease Prevention and Weight Management by J.B. Anderson, Ph.D. & Marilyn C. Sparling, MA, MPH, RD

Cooked by Michael Pollan

PRODUCTS

Extra-large Pan Pro Non-Stick Baking Sheets from Crate & Barrel. A great sheet pan that fills your whole oven! The secret to roasting vegetables it to not have them touch each other, so you want more surface.

Stainless Steel Apple Wedger from Pampered Chef. Steel instead of plastic.

24ct Silicone Mini Muffin Pan Made By Design™ at Target and **12ct Silicone Muffin Pan** Made By Design™ at Target. These pans aren't floppy, so you can pick them up with one hand, and the muffins pop right out by pushing up from underneath after the pan has cooled a bit.

Electric Egg Cooker by Chef's Choice. Cooks eggs to your desired doneness while doing something else. You can hear the buzzer from another room.

Thermapen Mk4 Thermometer by ThermoWorks. Using a thermometer takes all the guesswork out of cooking fish, poultry, pork, and beef. This particular thermometer is amazing: it works super-fast and is waterproof.

OXO Complete Grate and Slice Set at Target. This cheese grater is easy to use, and the lid doubles as a grating surface and a food container.

Onion Chopper Pro Deluxe Blade Set by Müeller. A strong, tear-reducing chopper that quickly and effortlessly chops onions.

Salad Cutting Bowl Set from Pampered Chef. Cut the vegetables all at once inside the bowl. The knife that comes with the set fits in the slats of the bowl and is an awesome knife for any produce.

CPSIA information can be obtained
at www.ICGtesting.com
Printed in the USA
BVHW090932140620
581454BV00004B/11